ALGEBRA
Prerequisite Skills Workbook: Remediation and Intervention

For use with
Glencoe Pre-Algebra
Glencoe Algebra 1
Glencoe Algebra: Concepts and Applications

New York, New York
Columbus, Ohio
Chicago, Illinois
Peoria, Illinois
Woodland Hills, California

Send all inquiries to:
Glencoe/McGraw-Hill
8787 Orion Place
Columbus, OH 43240-4027

ISBN: 0-07-827759-0

Algebra Prerequisite Skills Workbook

2 3 4 5 6 7 8 9 10 024 11 10 09 08 07 06 05 04 03

Contents

A. Whole Numbers

1. Comparing and Ordering Whole Numbers 1
2. Rounding Whole Numbers 3
3. Adding Whole Numbers 5
4. Subtracting Whole Numbers 7
5. Multiplying Whole Numbers 9
6. Dividing Whole Numbers 11
7. Divisibility Rules 13

B. Decimals

8. Decimals and Place Value 15
9. Rounding Decimals 17
10. Comparing and Ordering Decimals ... 19
11. Adding Decimals 21
12. Subtracting Decimals 23
13. Multiplying Decimals by Whole Numbers 25
14. Multiplying Decimals by Decimals 27
15. Dividing Decimals by Whole Numbers 29
16. Dividing Decimals by Decimals 31
17. Multiplying Decimals by Powers of Ten 33
18. Dividing Decimals by Powers of Ten 35

C. Fractions and Mixed Numbers

19. Equivalent Fractions 37
20. Simplifying Fractions 39
21. Writing Improper Fractions as Mixed Numbers 41
22. Writing Mixed Numbers as Improper Fractions 43
23. Comparing and Ordering Fractions 45
24. Multiplying Fractions 47
25. Multiplying Fractions and Mixed Numbers 49
26. Dividing Fractions 51
27. Dividing Fractions and Mixed Numbers 53
28. Adding Fractions 55
29. Adding Fractions and Mixed Numbers 57
30. Subtracting Fractions 59
31. Subtracting Fractions and Mixed Numbers 61

D. Fractions, Decimals, and Percents

32. Writing Fractions as Decimals 63
33. Writing Decimals as Fractions 65
34. Writing Decimals as Percents 67
35. Writing Percents as Decimals 69
36. Writing Fractions as Percents 71
37. Writing Percents as Fractions 73
38. Comparing and Ordering Rational Numbers 75

E. Measurement

39. Length in the Customary System 77
40. Capacity in the Customary System ... 79
41. Weight in the Customary System 81
42. Length in the Metric System 83
43. Capacity in the Metric System 85
44. Mass in the Metric System 87
45. Converting Customary Units to Metric Units 89
46. Converting Metric Units to Customary Units 91
47. Adding and Converting Units of Time 93

F. Probability and Statistics

48. Line Graphs 95
49. Histograms 97
50. Probability 99

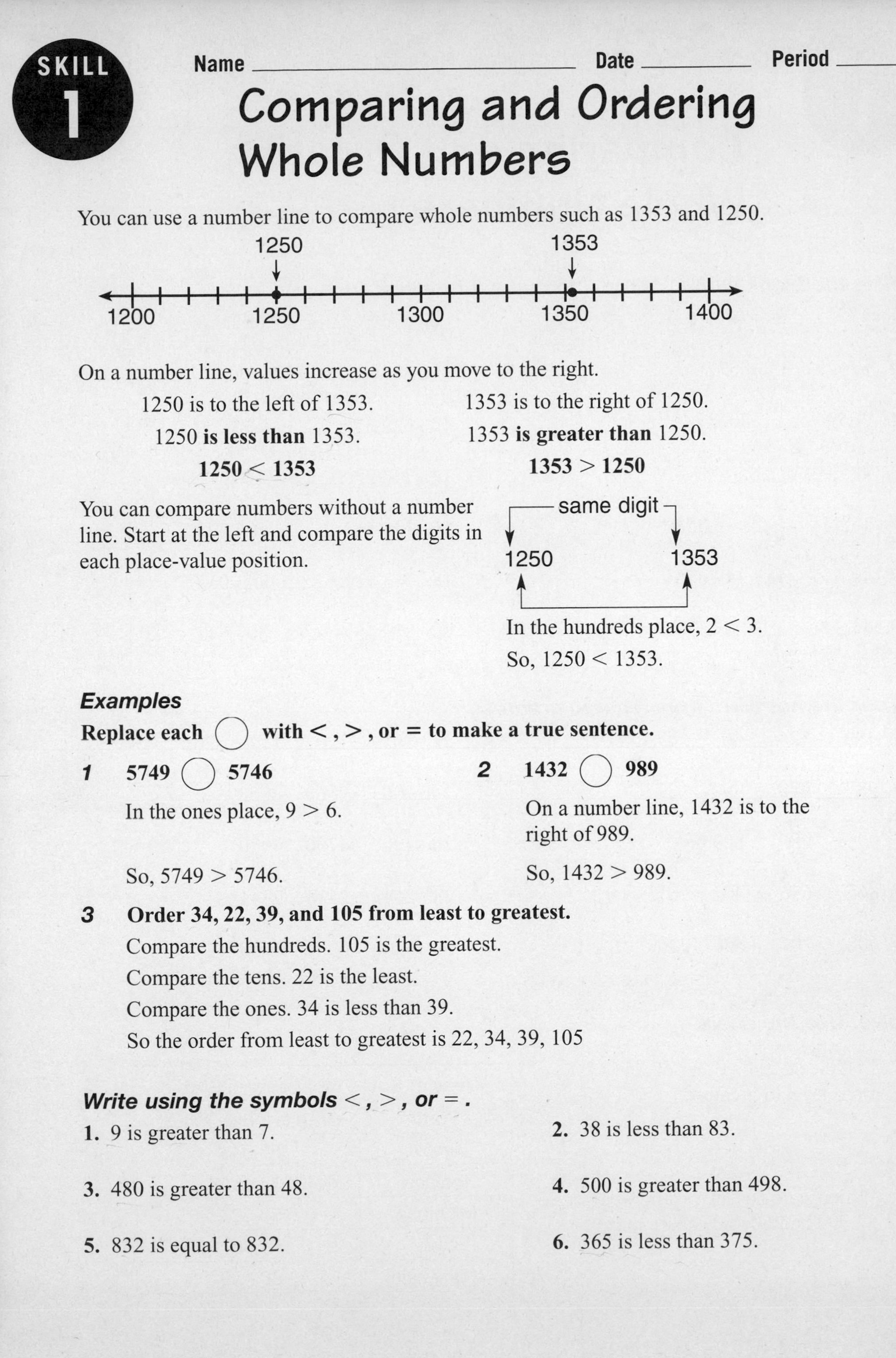

SKILL 1

Name ______________________ Date __________ Period ______

Comparing and Ordering Whole Numbers

You can use a number line to compare whole numbers such as 1353 and 1250.

On a number line, values increase as you move to the right.

1250 is to the left of 1353. 1353 is to the right of 1250.

1250 **is less than** 1353. 1353 **is greater than** 1250.

$\mathbf{1250 < 1353}$ $\mathbf{1353 > 1250}$

You can compare numbers without a number line. Start at the left and compare the digits in each place-value position.

In the hundreds place, $2 < 3$.

So, $1250 < 1353$.

Examples

Replace each ◯ with $<$, $>$, or $=$ to make a true sentence.

1 **5749 ◯ 5746**

In the ones place, $9 > 6$.

So, $5749 > 5746$.

2 **1432 ◯ 989**

On a number line, 1432 is to the right of 989.

So, $1432 > 989$.

3 **Order 34, 22, 39, and 105 from least to greatest.**

Compare the hundreds. 105 is the greatest.

Compare the tens. 22 is the least.

Compare the ones. 34 is less than 39.

So the order from least to greatest is 22, 34, 39, 105

Write using the symbols $<$, $>$, or $=$.

1. 9 is greater than 7.

2. 38 is less than 83.

3. 480 is greater than 48.

4. 500 is greater than 498.

5. 832 is equal to 832.

6. 365 is less than 375.

SKILL 1

Name ____________________ Date __________ Period ______

Comparing and Ordering Whole Numbers *(continued)*

Fill in the blank with <, >, or = to make a true sentence.

7. 435 _____ 534

8. 6739 _____ 6738

9. 8762 _____ 8672

10. 892 _____ 2531

11. 7059 _____ 7061

12. 629,356 _____ 630,200

13. 487,926 _____ 487,826

14. 74,923 _____ 74,923

15. 15,538 _____ 15,358

16. 124,462 _____ 124,433

17. 49,675 _____ 49,675

18. 753,021 _____ 743,012

19. 64,336 _____ 65,376

20. 819,461 _____ 803,642

Order the numbers from least to greatest.

21. 48 52 46 67

22. 102 120 112 201

23. 987 978 990 897

24. 2063 2060 2058

25. 99 989 809

26. 4007 4700 4070

27. 865 635 402 615

28. 2143 2413 2341

29. 602 206 620 260

30. 6300 6003 6030

Solve. Use the chart.

31. List the states in order of size from least to greatest.

32. Which state has an area between 57,000 and 60,000 square miles?

Areas of Some Midwestern States

State	Area (square miles)
Illinois	56,345
Indiana	36,185
Michigan	58,527
Ohio	41,330
Wisconsin	56,123

SKILL 2

Name ______________________ Date __________ Period ______

Rounding Whole Numbers

The distance from Atlanta, Georgia, to Memphis, Tennessee, is 371 miles. If you tell a friend that the distance is about 400 miles, you have **rounded** the number.

On a number line, you can see that 371 is between 300 and 400. It is closer to 400. *To the nearest hundred*, 371 rounds to 400.

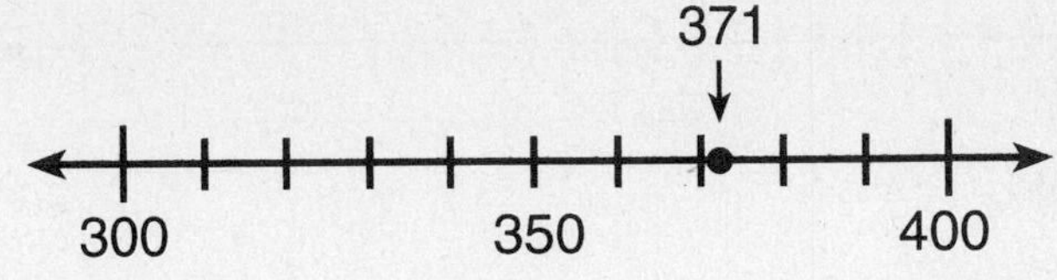

You can also round numbers without using a number line. First, look at the digit to the right of the place being rounded.

- If the digit to the right is 5, 6, 7, 8, or 9, round up.
- If the digit to the right is 0, 1, 2, 3, or 4, the underlined digit remains the same.

Examples

1 **Round 84,373 to the nearest thousand.**

8<u>4</u>,373

thousands ↗ *The digit in the thousand place remains the same since the digit to its right is 3.*

To the nearest thousand, 84,373 rounds to 84,000.

2 **Round 3,546,238 to the nearest million.**

<u>3</u>,546,238

millions ↗ *Round up since the digit is 5.*

To the nearest million, 3,546,238 rounds to 4,000,000.

Round to the nearest ten. Use the number line if necessary.

1. 682 **2.** 675 **3.** 698 **4.** 661

Round to the nearest hundred. Use the number line if necessary.

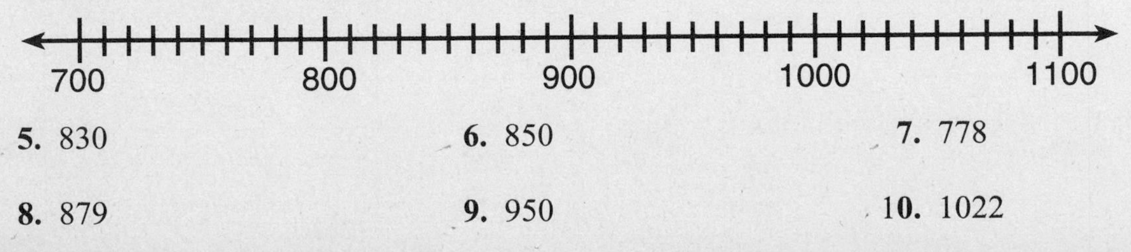

5. 830 **6.** 850 **7.** 778

8. 879 **9.** 950 **10.** 1022

Name ______________________ Date ____________ Period ________

Rounding Whole Numbers *(continued)*

Round to the nearest thousand. Use the number line if necessary.

2000 3000 4000 5000 6000

11. 3100 **12.** 2500 **13.** 2262

14. 4700 **15.** 5860 **16.** 4082

17. 3643 **18.** 4216 **19.** 5910

Round to the underlined place-value position.

20. 2$\underline{6}$7

21. 40$\underline{9}$1

22. $\underline{4}$20,800

23. 5$\underline{6}$7,000

24. 43,$\underline{7}$28

25. 30$\underline{7}$,792

26. 1$\underline{4}$,350

27. $\underline{9}$,798

28. $\underline{3}$,398,000

29. 1$\underline{8}$,499,898

30. 5$\underline{3}$2,795

31. $\underline{8}$24,619

32. $\underline{6}$,321,510

33. 2$\underline{4}$,053,217

34. 12$\underline{7}$,610,573

35. 34$\underline{6}$,872,000

Solve. Use the chart.

36. List the oceans in order of size from least area to greatest area.

37. Round each area to the nearest million.

Areas of Oceans	
Ocean	**Area (square kilometers)**
Arctic	9,485,000
Atlantic	86,557,000
Indian	73,427,000
Pacific	166,241,000

Name ______________________ Date ____________ Period ________

Adding Whole Numbers

To add whole numbers, first add the ones. Then add the digits in each place from right to left.

Examples

1

```
 7056   →     ¹           →     ¹ ¹         →     ¹ ¹
+ 973   →    7056         →    7056         →    7056
-----       + 973              + 973              + 973
    9       -----              -----              -----
               29                029               8029
```

Add the ones. *Add the tens.* *Add the hundreds.* *Add the thousands.*

2 $406 + $881 + $75

```
   ¹ ¹
  $406
   881
+   75
------
 $1362
```

Write in columns.

Add.

1. 72 + 65

2. 62 + 83

3. 39 + 37

4. 66 + 85

5. 768 + 67

6. 495 + 48

7. $470 + 583

8. 237 + 579

9. 1570 + 2823

10. 5126 + 2899

11. 3973 + 1689

12. 1482 + 3497

SKILL 3

Name ______________________ Date __________ Period ______

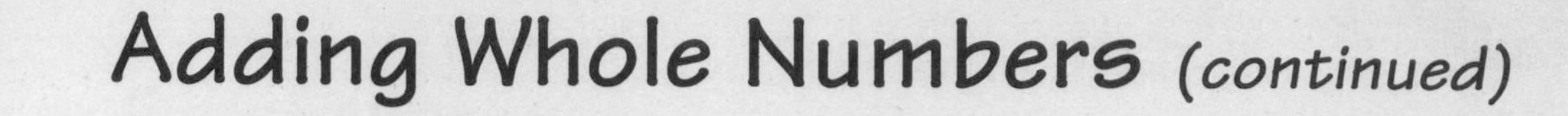

Adding Whole Numbers (continued)

13. $\begin{array}{r} 4632 \\ +\ 705 \\ \hline \end{array}$

14. $\begin{array}{r} 2039 \\ +\ 758 \\ \hline \end{array}$

15. $\begin{array}{r} 6720 \\ +\ 2385 \\ \hline \end{array}$

16. $\begin{array}{r} 7916 \\ +\ 8295 \\ \hline \end{array}$

17. $\begin{array}{r} 14{,}832 \\ +\ 6229 \\ \hline \end{array}$

18. $\begin{array}{r} 23{,}467 \\ +\ 7324 \\ \hline \end{array}$

19. $\begin{array}{r} 15{,}732 \\ +\ 8615 \\ \hline \end{array}$

20. $\begin{array}{r} 24{,}816 \\ +\ 15{,}995 \\ \hline \end{array}$

21. $\begin{array}{r} 36 \\ 54 \\ +\ 21 \\ \hline \end{array}$

22. $\begin{array}{r} 65 \\ 89 \\ +\ 23 \\ \hline \end{array}$

23. $\begin{array}{r} 168 \\ 275 \\ +\ 256 \\ \hline \end{array}$

24. $\begin{array}{r} 245 \\ 87 \\ +\ 316 \\ \hline \end{array}$

25. $\begin{array}{r} 43 \\ 128 \\ +\ 210 \\ \hline \end{array}$

26. $\begin{array}{r} 439 \\ 64 \\ +\ 87 \\ \hline \end{array}$

27. $\begin{array}{r} 518 \\ 192 \\ +\ 36 \\ \hline \end{array}$

28. $\begin{array}{r} 425 \\ 376 \\ +\ 124 \\ \hline \end{array}$

29. 5 + 27 + 168 =

30. 463 + 309 + 1542 =

31. $46 + $93 + $18 + $62 =

32. 636 + 4923 + 481 =

Solve.

33. Karen had $273 in her savings account. She makes deposits of $15 and $43. How much does Karen have in her savings account now?

34. One day, 148 copies of the student newspaper were sold. On the previous day, 164 copies were sold. How many copies were sold during these two days?

Name ______________________ Date ____________ Period ________

Subtracting Whole Numbers

To subtract whole numbers, first subtract the ones. Then subtract the digits in each place from right to left. Rename as needed.

Examples

1

$$\begin{array}{r} 896 \\ -\ 145 \\ \hline 1 \end{array} \longrightarrow \begin{array}{r} 896 \\ -\ 145 \\ \hline 51 \end{array} \longrightarrow \begin{array}{r} 896 \\ -\ 145 \\ \hline 751 \end{array}$$

Subtract the ones. *Subtract the tens.* *Subtract the hundreds.*

2

$$\begin{array}{r} 381 \\ -\ 285 \\ \hline \end{array} \longrightarrow \begin{array}{r} 381 \\ -\ 285 \\ \hline 6 \end{array} \longrightarrow \begin{array}{r} 381 \\ -\ 285 \\ \hline 96 \end{array}$$

(renamed: 7 11 above the 8 and 1; then 2 17 11 above 3, 8 and 1)

Since 1 < 5, rename 8 tens as 7 tens and 10 ones. Then, 10 ones + 1 one = 11 ones.

3

$$\begin{array}{r} 506 \\ -\ 238 \\ \hline \end{array} \quad \begin{array}{r} 506 \\ -\ 238 \\ \hline 8 \end{array} \quad \begin{array}{r} 506 \\ -\ 238 \\ \hline 268 \end{array}$$

(renamed: 49 above 50, 16 above 6)

Since 6 < 8, rename 50 tens as 49 tens 10 ones. Then, 10 ones + 6 ones = 16 ones.

Subtract.

1. $\begin{array}{r} 87 \\ -\ 53 \\ \hline \end{array}$ **2.** $\begin{array}{r} 56 \\ -\ 40 \\ \hline \end{array}$ **3.** $\begin{array}{r} 854 \\ -\ 630 \\ \hline \end{array}$ **4.** $\begin{array}{r} 695 \\ -\ 132 \\ \hline \end{array}$

5. $\begin{array}{r} 34 \\ -\ 8 \\ \hline \end{array}$ **6.** $\begin{array}{r} 70 \\ -\ 28 \\ \hline \end{array}$ **7.** $\begin{array}{r} \$78 \\ -\ 59 \\ \hline \end{array}$ **8.** $\begin{array}{r} 480 \\ -\ 63 \\ \hline \end{array}$

SKILL 4

Name ________________ Date __________ Period ______

Subtracting Whole Numbers *(continued)*

9. 407 − 139

10. 908 − 439

11. 320 − 152

12. 300 − 105

13. 515 − 298

14. 735 − 596

15. 810 − 635

16. 401 − 293

17. 6827 − 5752

18. 1297 − 898

19. 6243 − 4564

20. 5690 − 792

21. 1516 − 835 =

22. 8312 − 5943 =

23. 16,202 − 9814 =

24. 12,915 − 8036 =

25. 51,520 − 35,630 =

26. 37,982 − 19,395 =

27. 70,605 − 38,296 =

28. 30,005 − 17,008 =

Solve.

29. A cassette recorder costs $340 at one store. At another store, the same brand costs $298. How much would you save by buying the recorder at the second store?

30. The Colorado River is 1,450 miles long. The Yukon River is 1,770 miles long. How much longer is the Yukon River?

SKILL 5

Name ______________________ Date __________ Period ______

Multiplying Whole Numbers

To multiply by a one-digit whole number, first multiply the ones. Then multiply the digits in each place from right to left.

Example

1

$$\begin{array}{r} \overset{3}{8}\,\overset{}{3}5 \\ \times\quad 6 \\ \hline 0 \end{array} \longrightarrow \begin{array}{r} \overset{2\,3}{835} \\ \times\quad 6 \\ \hline 10 \end{array} \longrightarrow \begin{array}{r} \overset{2\,3}{835} \\ \times\quad 6 \\ \hline 5010 \end{array}$$

Multiply the ones. — *Multiply the tens. Add 3.* — *Multiply the hundreds. Add 2.*

To multiply by a two digit whole number, first multiply by the ones. Then multiply by the tens.

Examples

2

$$\begin{array}{r} 2609 \\ \times\quad 78 \\ \hline \end{array} \longrightarrow \begin{array}{r} 2609 \\ \times\quad 78 \\ \hline 20872 \end{array} \longrightarrow \begin{array}{r} 2609 \\ \times\quad 78 \\ \hline 20872 \\ 182630 \\ \hline 203{,}502 \end{array}$$

3

$$\begin{array}{r} 1047 \\ \times\quad 60 \\ \hline \end{array} \longrightarrow \begin{array}{r} 1407 \\ \times\quad 60 \\ \hline 0 \end{array} \longrightarrow \begin{array}{r} 1407 \\ \times\quad 60 \\ \hline 62{,}820 \end{array}$$

Any number multiplied by zero is zero.

Multiply.

1. $\begin{array}{r} 700 \\ \times\ 25 \\ \hline \end{array}$ **2.** $\begin{array}{r} 602 \\ \times\ 4 \\ \hline \end{array}$ **3.** $\begin{array}{r} 218 \\ \times\ 63 \\ \hline \end{array}$ **4.** $\begin{array}{r} \$189 \\ \times\ 42 \\ \hline \end{array}$

5. $\begin{array}{r} \$125 \\ \times\ 11 \\ \hline \end{array}$ **6.** $\begin{array}{r} 264 \\ \times\ 40 \\ \hline \end{array}$ **7.** $\begin{array}{r} 3265 \\ \times\ 72 \\ \hline \end{array}$ **8.** $\begin{array}{r} 6019 \\ \times\ 94 \\ \hline \end{array}$

SKILL 5

Name ______________________ Date ____________ Period ______

Multiplying Whole Numbers *(continued)*

9. $\begin{array}{r} 3841 \\ \times \quad 65 \\ \hline \end{array}$

10. $\begin{array}{r} \$7903 \\ \times \quad 3 \\ \hline \end{array}$

11. $\begin{array}{r} 16{,}009 \\ \times \quad 80 \\ \hline \end{array}$

12. $\begin{array}{r} 28{,}706 \\ \times \quad 49 \\ \hline \end{array}$

13. $\begin{array}{r} 4216 \\ \times \quad 8 \\ \hline \end{array}$

14. $\begin{array}{r} 5310 \\ \times \quad 50 \\ \hline \end{array}$

15. $\begin{array}{r} 8020 \\ \times \quad 16 \\ \hline \end{array}$

16. $\begin{array}{r} 19{,}634 \\ \times \quad 25 \\ \hline \end{array}$

17. $819 \times 8 =$

18. $438 \times 6 =$

19. $6420 \times 40 =$

20. $7253 \times 38 =$

21. $\$8053 \times 5 =$

22. $450 \times 30 =$

23. $\$605 \times 15 =$

24. $79{,}025 \times 61 =$

Solve.

25. There are 42 rows of seats in the theater. There are 36 seats in each row. How many seats are in the theater?

26. A truck carries 278 crates. Each crate holds 45 pounds of fruit. How many pounds of fruit does the truck carry?

Name ______________________ Date __________ Period

Dividing Whole Numbers

To divide whole numbers, start with the digit in the left most position. Then divide the digit in each place from left to right.

Examples

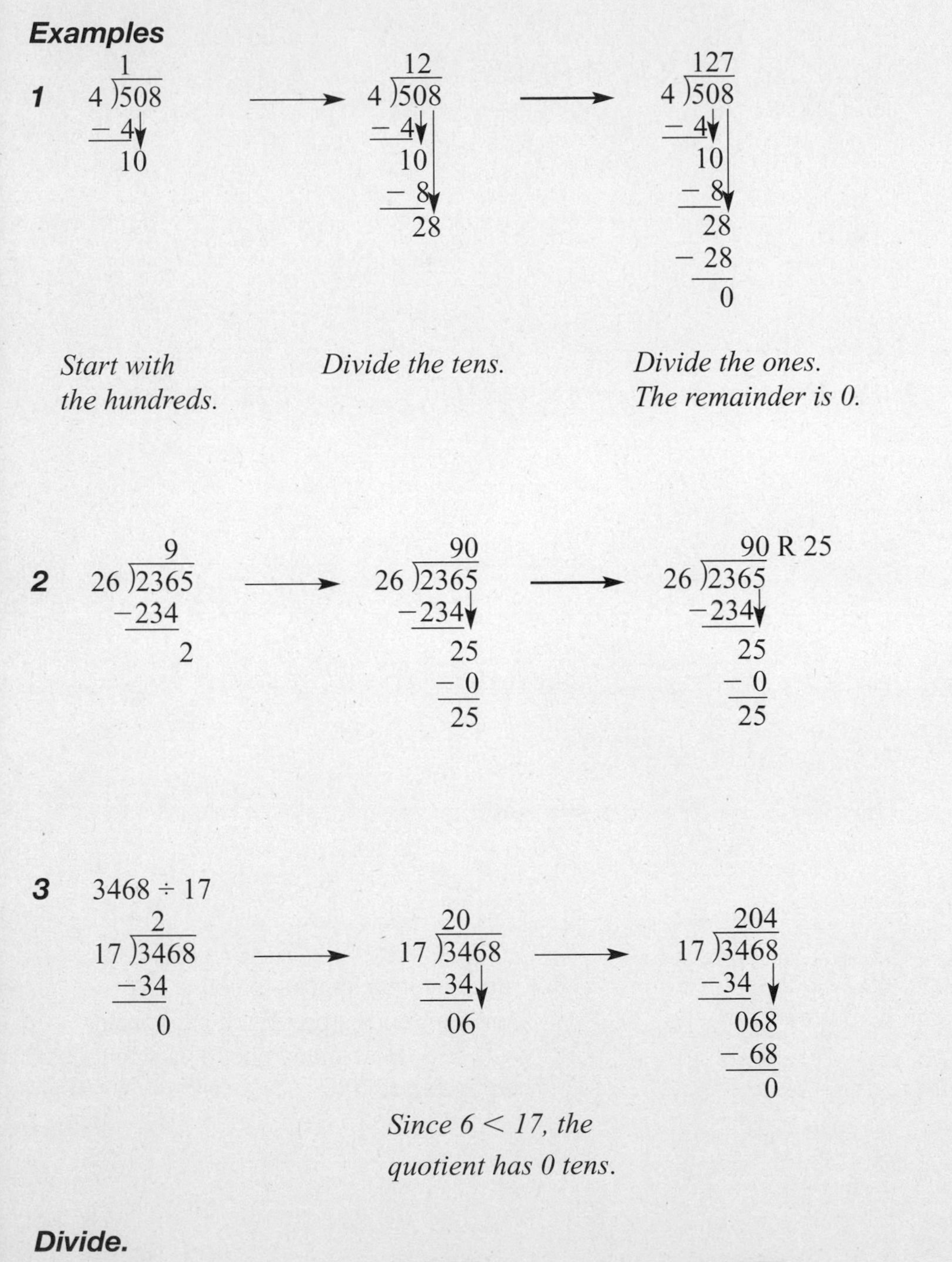

Divide.

1. 5)3255 **2.** 70)359 **3.** 47)517 **4.** 18)901

SKILL 6

Name ______________________ Date __________ Period ______

Dividing Whole Numbers *(continued)*

Divide.

5. $65\overline{)1300}$
6. $50\overline{)2500}$
7. $59\overline{)3776}$
8. $23\overline{)1187}$
9. $15\overline{)1260}$
10. $9\overline{)769}$
11. $6\overline{)5246}$
12. $12\overline{)1176}$
13. $27\overline{)1435}$
14. $37\overline{)592}$
15. $37\overline{)1000}$
16. $81\overline{)5430}$
17. $46\overline{)\$1656}$
18. $42\overline{)2480}$
19. $86\overline{)3440}$
20. $62\overline{)1858}$
21. $72\overline{)43,704}$
22. $46\overline{)20,700}$
23. $5202 \div 18 =$
24. $2619 \div 3 =$
25. $37,513 \div 4 =$
26. $4886 \div 17 =$

Solve.

27. Each tent is put up with 12 poles. How many tents can be put up with 200 poles?

18. Gary buys backpacks to sell at his sporting goods store. Each backpack costs $38. How many backpacks can he buy for $270?

SKILL 7

Name ________________________ Date __________ Period 

Divisibility Rules

The following rules will help you determine if a number is divisible by 2, 3, 4, 5, 6, 8, 9, or 10.

A number is divisible by:

- 2 if the ones digit is divisible by 2.
- 3 if the sum of the digits is divisible by 3.
- 4 if the number formed by the last two digits is divisible by 4.
- 5 if the ones digit is 0 or 5.
- 6 if the number is divisible by 2 and 3.
- 8 if the number formed by the last three digits is divisible by 8.
- 9 if the sum of the digits is divisible by 9.
- 10 if the ones digit is 0.

Example **Determine whether 2120 is divisible by 2, 3, 4, 5, 6, 9, or 10.**

2: The ones digit is divisible by 2.
2120 is divisible by 2.

3: The sum of the digits $2 + 1 + 2 + 0 = 5$, is not divisible by 3.
2120 is not divisible by 3.

4: The number formed by the last two digits, 20, is divisible by 4.
2120 is divisible by 4.

5: The ones digit is 0.
2120 is divisible by 5.

6: The number is divisible by 2 but not by 3.
2120 is not divisible by 6.

8: The number formed by the last 3 digits, 120, is divisible by 8.
2120 is divisible by 8.

9: The sum of the digits, $2 + 1 + 2 + 0 = 5$, is not divisible by 9.
2120 is not divisible by 9.

10: The ones digit is 0.
2120 is divisible by 10.

2120 is divisible by 2, 4, 5, 8, and 10.

Determine whether the first number is divisible by the second number. Write yes or no.

1. 4829; 9
2. 482; 2
3. 1692; 6
4. 1355; 10
5. 633; 3
6. 724; 4
7. 3714; 8
8. 912; 9
9. 559; 5
10. 20,454; 6
11. 616; 8
12. 3000; 4

SKILL 7

Name ______________________ Date ____________ Period ________

Divisibility Rules *(continued)*

Determine whether each number is divisible by 2, 3, 4, 5, 6, 8, 9, or 10.

13. 80

14. 91

15. 180

16. 333

17. 1024

18. 11,010

19. Is 9 a factor of 154?

20. Is 6 a factor of 102?

21. Is 486 divisible by 6?

22. Is 441 divisible by 9?

Determine whether the first number is divisible by the second number.

23. 4281; 2

24. 2670; 10

25. 3945; 6

26. 6132; 4

27. 8304; 3

28. 6201; 9

29. 4517; 9

30. 2304; 8

31. 7000; 5

32. 10,000; 8

33. 9420; 6

34. 822; 4

Use mental math to find a number that satisfies the given conditions.

35. a number divisible by both 3 and 5

36. a four-digit number divisible by 3, but *not* by 9

37. a five-digit number *not* divisible by 3 or 10

38. a four-digit number divisible by 2 and 4, but *not* by 8

Name ______________________ Date __________ Period ______

Decimals and Place Value

You can use a place-value chart like the one below to help you write and read decimals and understand their values.

The decimal 160.289 is shown in the chart at the right. The place-value chart can be extended in either direction. The digit 9, together with its place value, names the number nine thousandths or 0.009.

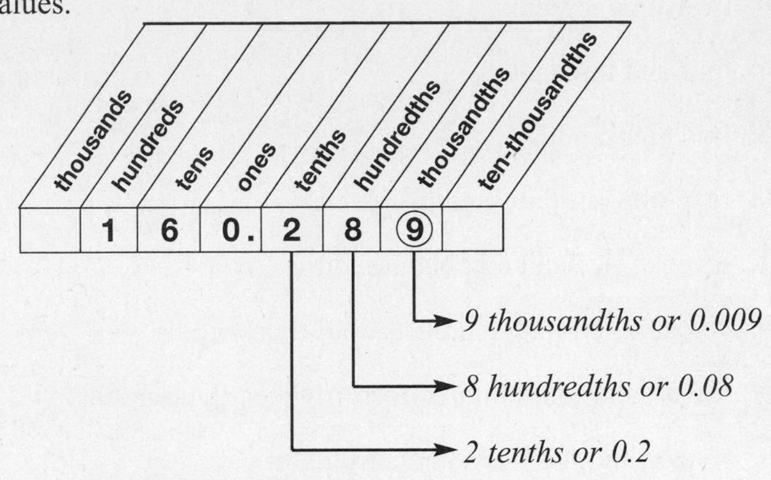

Notice that the decimal point separates the ones and tenths places. It is read as *and.*

The decimal 160.289 is read as *one hundred sixty and two hundred eighty-nine thousandths.*

Examples ***1*** **Write nine and five hundred twenty-six ten-thousandths as a number.**

9.0526

2 **Write 623.75 in words.**

six hundred twenty-three and seventy-five hundredths

Write the number named by the underlined digit in words.

1. $0.\underline{4}5$ **2.** $2.36\underline{9}$ **3.** $110.5\underline{1}$

4. $43.\underline{6}72$ **5.** $98.00\underline{8}$ **6.** $5.312\underline{6}$

7. $16.0\underline{9}$ **8.** $2.06\underline{7}4$ **9.** $2.067\underline{4}$

10. $0.0\underline{8}7$ **11.** $0.0\underline{2}51$ **12.** $7.585\underline{7}$

Name ____________________ Date __________ Period ______

Decimals and Place Value *(continued)*

Write each of the following as a decimal.

13. twelve hundredths

14. four and three tenths

15. five thousandths

16. fifty-one ten-thousandths

17. seventy-five and nine thousandths

18. one hundred four and thirty-four thousandths

19. twenty and four hundred forty-five ten-thousandths

20. sixteen and forty-five thousandths

21. fifty-six and thirty-four hundredths

Write each number in words.

22. 6.04

23. 0.017

24. 5.1648

25. 18.456

26. 145.007

27. 28.796

28. 787.462

29. 9.0045

In the 1996 Olympics, Michael Johnson won both the men's 200-meter and 400-meter track competitions.

30. His time for the 200-meter competition was 19.32 seconds Write this decimal in words.

31. His time for the 400-meter competition was forty-three and forty–nine hundredths seconds. Write this as a decimal.

Name ______________________ Date ____________ Period ______

Rounding Decimals

Round 34.725 to the nearest tenth.

You can use a number line.

Find the approximate location of 34.725 on the number line.

34.725 is closer to 34.7 than to 34.8
34.725 rounded to the nearest tenth is 34.7.

34.0 34.1 34.2 34.3 34.4 34.5 34.6 34.7 34.8 34.9 35.0

You can also round without a number line.

Find the place to which you want to round.	Look at the digit to the right. If the digit is less than 5, round down. If the digit is 5 or greater, round up.	2 is less than 5. Round down.
34.725	34.7**2**5	34.7

Use each number line to show how to round the decimal to the nearest tenth.

1. 7.82 — 7.0 7.1 7.2 7.3 7.4 7.5 7.6 7.7 7.8 7.8 8.0

2. 0.39 — 0.0 0.1 0.2 0.3 0.4 0.5 0.6 0.7 0.8 0.9 1.0

3. 5.071 — 5.0 5.1 5.2 5.3 5.4 5.5 5.6 5.7 5.8 5.9 6.0

Round each number to the underlined place-value position.

4. $6.\underline{3}2$ **5.** $0.4\underline{7}21$ **6.** $26.\underline{4}44$ **7.** $1.\underline{1}61$

8. $362.08\underline{4}6$ **9.** $15.5\underline{5}3$ **10.** $151.3\underline{9}1$ **11.** $0.\underline{5}5$

12. $631.00\underline{0}8$ **13.** $17.3\underline{2}7$ **14.** $3.\underline{0}9$ **15.** $1.\underline{5}8$

SKILL 9

Name ____________________ Date ____________ Period ________

Rounding Decimals *(continued)*

Round each number to the underlined place-value position.

16. $1.7\underline{2}6$

17. $5\underline{4}.38$

18. $0.\underline{5}8$

19. $0.9\underline{1}42$

20. $80.\underline{6}59$

21. $23\underline{2}.1$

22. $1.\underline{0}63$

23. $0.\underline{5}5$

24. $0.\underline{8}194$

25. $0.4\underline{9}6$

26. $3.01\underline{8}2$

27. $71.\underline{4}05$

28. $\underline{9}.63$

29. $32.\underline{7}1$

30. $2.6\underline{7}1$

31. $4.05\underline{0}7$

32. $89.\underline{9}5$

33. $0.1\underline{3}4$

34. $5.\underline{8}93$

35. $52\underline{0}.6$

36. $0.70\underline{9}8$

37. $1.8\underline{4}5$

38. $34.\underline{5}5$

39. $29.\underline{2}5$

40. $56.09\underline{2}4$

41. $119\underline{9}.7$

42. $0.\underline{4}6$

43. $0.\underline{3}546$

Name ______________________ Date __________ Period ______

Comparing and Ordering Decimals

To compare decimals, you compare digits in each place-value position from left to right.

Examples **1** **Compare 3.0752 and 3.1042.**

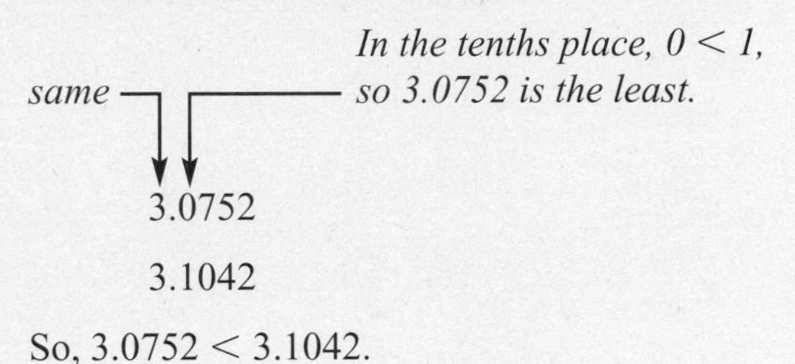

So, 3.0752 < 3.1042.

2 **Fill in the blank with <, >, or = to make a true sentence.**

14.19 ________ 14.103

In the hundredths place, 9 > 0. So 14.19 > 14.103.

3 **Order the following set of decimals from least to greatest.**

8.4, 8.41, 8.406, 8.442

Annex zeros so all decimals have the same number of place-value positions.

8.400, 8.410, 8.406, 8.442

So, 8.400 < 8.406 < 8.410 < 8.442.

The decimals in order from least to greatest are

8.4, 8.406, 8.41, 8.442.

State whether each statement is true or false.

1. 0.3 = 0.30

2. 0.001 = 0.01

3. 0.7 > 0.8

4. 0.204 < 0.24

5. 17 = 17.00

6. 0.9 > 2.0

SKILL 10

Name ______________________ Date __________ Period ______

Comparing and Ordering Decimals *(continued)*

Fill in the blank with $<$, $>$, or $=$ to make a true sentence.

7. 0.205 _____ 0.250

8. 6.035 _____ 6.0353

9. 0.40 _____ 0.400

10. 0.55 _____ 0.5

11. 6.4 _____ 6.400

12. 1.05 _____ 1.005

13. 0.002 _____ 0.02

14. 0.615 _____ 0.651

15. 7 _____ 7.00

16. 15.3 _____ 15.30

17. 11.01 _____ 11.10

18. 124.6 _____ 124.48

Order each set of decimals from least to greatest.

19. 0.03, 0.3, 0.003, 3.0

20. 5.23, 5.203, 5.21, 5.3

21. 0.91, 0.866, 0.9, 0.87

22. 2.03, 2.13, 2.3, 2.033

23. 16.4, 16.04, 16.45, 16.001

24. 8.7, 8.07, 8.17, 8.01

25. 114.2, 114.02, 114.202, 114.002

26. 0.362, 0.306, 0.31, 0.36

Solve.

27. In gymnastics, Maria receives an average score of 9.7. Rebecca receives an average score of 9.69. Who is the winner?

28. Three golfers have the following stroke averages. Rank the golfers in order from lowest to highest stroke average.

Lopez	71.2
Higuchi	72.17
Blalock	72.15

SKILL 11

Name ______________________ Date ____________ Period ________

Adding Decimals

To add decimals, first line up the decimal points. Then add as with whole numbers.

Examples ***1*** **Add: 36.801 + 8.945.**

```
 11
 36.801
+ 8.945
-------
 45.746
```

2 **Add: 7.3 + 9 + 8.45.**

```
  7.30
  9.00      Write 9 as 9.00.
+ 8.45
------
 24.75
```

3 **Add: $415 + $29.05.**

```
   1
$415.00     Annex zeros to $415 to help align
+ 29.05     the decimal points.
-------
$444.05
```

Add.

1. $27.06 + 7.06

2. 1.034 + 0.08

3. 68.7 + 8.41

4. 42.6 + 21.919

5. 93.7 + 24.85

6. 140.98 + 16.5

7. 15.987 + 9.07

8. 478.98 + 99.076

9. 14.16 + 8.9

10. 67.032 + 5.98

11. 246.38 + 19.976

12. 17.32 + 8.963

Name ______________________ Date ____________ Period ______

Adding Decimals *(continued)*

Add.

13. $510.35 + 6.7$

14. $83.675 + 2.95$

15. $6.852 + 3.97$

16. $14.8 + 9.63$

17. $0.4 + 0.6 + 0.7$

18. $6.5 + 2.81 + 7.9$

19. $0.21 + 0.619 + 0.394$

20. $\$3.33 + 6.67 + 0.24$

21. $7.41 + 2.835 + 0.9$

22. $\$19.99 + 7.99 + 24.50$

23. $3.04 + 0.6 =$

24. $8 + 4.7 =$

25. $19.642 + 2.61 =$

26. $8.543 + 3.29 =$

27. $1.61 + 3.807 =$

28. $543 + 9.29 =$

Solve.

29. A gymnast scored 9.65 on the beam, 9.59 on the floor, 9.76 on the bars, and 9.52 on the vault. What was the gymnast's total score?

30. A ticket to the game cost Andrea $12. She also spent $8.09 on food. How much did she spend in all?

Name ______________________ Date __________ Period ______

Subtracting Decimals

To subtract decimals, line up the decimal points.
Then subtract as with whole numbers.

Examples ***1*** **Subtract: 8.1 − 4.75.**

```
  0 10
 8.1̸0̸
− 4.75
 3.35
```

Annex a zero to 8.1 to help align the decimal points.

2 **Subtract: \$84 − \$1.79.**

```
  3 9 10
$84̸.0̸0̸
−  1.79
$82.21
```

Annex two zeros to \$84 to help align the decimal points.

3 **Subtract: 16.703 − 8.**

```
 16.703
− 8.000
  8.703
```

Annex three zeros to 8 to help align the decimal points.

Subtract.

1. 9.14 − 2.075

2. 712.53 − 6.44

3. 20.14 − 8.093

4. \$12.65 − 10.99

5. 14.395 − 2.654

6. 2.42 − 0.5

7. 0.261 − 0.09

8. 9.407 − 0.22

9. 6.324 − 0.75

10. 42.903 − 8.05

11. 16.37 − 5.609

12. 18 − 7.63

Name ______________________ Date __________ Period ______

Subtracting Decimals *(continued)*

Subtract.

13. $142.6 - 85.92$

14. $25.37 - 8.889$

15. $48.3 - 6.75$

16. $237.84 - 6.964$

17. $581.2 - 106.81$

18. $99.2 - 38.576$

19. $12.752 - 6.9$

20. $639.07 - 64.961$

21. $4 - 1.5$

22. $0.4 - 0.15$

23. $112.8 - 81.93$

24. $\$26 - 0.81$

25. $1 - 0.37$

26. $14.9 - 8.261$

27. $\$73 - 9.69$

28. $5 - 0.088$

29. $6.51 - 0.8 =$

30. $10.86 - 6.872 =$

31. $2.43 - 0.965 =$

32. $\$81 - \$4.83 =$

33. $210 - 56.765 =$

34. $16.7 - 0.082 =$

Solve.

35. Mrs. Taylor's class has earned $190.32 for their class project. They need $250. How much more do they need to earn?

36. Connie has 20 mL of sulfuric acid. Her experiment calls for 1.6 mL. How many mL will Connie have left after the experiment?

Name ______________________ Date __________ Period ______

Multiplying Decimals by Whole Numbers

To multiply a decimal by a whole number, first multiply as with whole numbers. Then place the decimal point in the product. The product has the same number of decimal places as the decimal factor.

Examples ***1*** **Multiply: 421 × 0.6.**

```
  421
× 0.6    ←——   1 decimal place in the decimal factor
-----
252.6    ←——   1 decimal place in the product
```

2 **Multiply: $6.16 × 47.**

```
  $6.16   ←——   2 decimal places in the decimal factor
 ×   47
 ------
   4312
  24640
-------
$289.52   ←——   2 decimal places in the product
```

Multiply.

1. 23 × 0.8

2. 45 × 0.9

3. 216 × 0.2

4. $0.83 × 7

5. $4.16 × 15

6. 27 × 0.6

7. 0.63 × 4

8. $5.65 × 14

9. 231 × 0.41

10. 0.62 × 11

11. $7.44 × 26

12. 218 × 0.54

SKILL 13

Name ______________________ Date ____________ Period ________

Multiplying Decimals by Whole Numbers *(continued)*

Multiply.

13. 113×0.6

14. 2.48×24

15. 15.48×19

16. 214.8×37

17. 438×0.85

18. 395×2.63

19. 87×0.8

20. 416×0.38

21. $25 \times 0.15 =$

22. $206 \times \$0.49 =$

23. $\$0.23 \times 15 =$

24. $0.47 \times 35 =$

25. $19 \times 0.19 =$

26. $419 \times 2.3 =$

27. $4.67 \times 15 =$

28. $0.842 \times 93 =$

29. $\$16.50 \times 12 =$

30. $143 \times 0.55 =$

Solve.

31. Turkey is on sale for $0.89 per pound. How much does William pay for a 14-pound turkey?

32. A clothing fabric factory needs 3.25 yards of fabric to make one skirt. How many yards are needed to make 2,000 skirts?

Name ______________________ Date ____________ Period ________

Multiplying Decimals by Decimals

Multiply decimals just like you multiply whole numbers. The number of decimal places in the product is equal to the sum of the number of decimal places in the factors.

Example **Multiply 0.038 and 0.17.**

$$\begin{array}{r} 0.038 \\ \times\ 0.17 \\ \hline 266 \\ 38\ \\ \hline 0.00646 \end{array}$$

0.038 ⟵ *three decimal places*
0.17 ⟵ *two decimal places*
0.00646 ⟵ *five decimal places*

The product is 0.00646.

Place the decimal point in each product.

1. $1.47 \times 6 = 882$

2. $0.9 \times 2.7 = 243$

3. $6.48 \times 2.4 = 15552$

Multiply.

4. $\begin{array}{r} 0.8 \\ \times\ 7 \\ \hline \end{array}$

5. $\begin{array}{r} 0.04 \\ \times\ 0.3 \\ \hline \end{array}$

6. $\begin{array}{r} 0.16 \\ \times\ 26 \\ \hline \end{array}$

7. $\begin{array}{r} 0.003 \\ \times\ 4.2 \\ \hline \end{array}$

8. 12.2×0.06

9. 0.0015×0.15

10. 1.9×2.2

11. 3.59×0.02

12. 12.2×0.007

13. 0.7×3.11

SKILL 14

Name ______________________ Date __________ Period ______

Multiplying Decimals by Decimals *(continued)*

Multiply.

14. $\begin{array}{r} 0.6 \\ \times\ 0.7 \\ \hline \end{array}$

15. $\begin{array}{r} 6.3 \\ \times\ 5.1 \\ \hline \end{array}$

16. $\begin{array}{r} 18.2 \\ \times\ 0.51 \\ \hline \end{array}$

17. 0.52×0.03

18. 0.29×29.1

19. 6.1×0.0054

20. 6.8×0.39

21. 3.57×0.09

22. 3.72×8.4

Solve each equation.

23. $t = 0.32 \times 0.05$

24. $6.4 \times 3.9 = h$

25. $k = 0.09 \times 2.3$

26. $a = 0.4 \times 9$

27. $0.23 \times 0.003 = m$

28. $1.09 \times 6.24 = v$

Evaluate each expression if m = 0.9 and n = 6.2.

29. $m \cdot 0.43$

30. $0.002 \cdot n$

31. $17.4 \cdot m$

Evaluate each expression if a = 0.4 and b = 5.8.

32. $0.48 \cdot a$

33. $b \cdot 13.8$

34. $0.003 \cdot a$

35. $1.4 \cdot b$

36. $3.6 \cdot a$

37. $24.5 \cdot a$

Name ______________________ Date __________ Period ______

Dividing Decimals by Whole Numbers

To divide a decimal by a whole number, first place the decimal point in the quotient directly above the decimal point in the dividend. Then divide as with numbers.

Examples ***1*** **Divide $58.10 by 7.**

```
     .                       8.                      $8.30
 7 )$58.10   ──→       7 )$58.10      ──→       7 )$58.10
                         − 56                     − 56↓ |
                           2                         2 1 |
                                                   − 2 1↓
                                                       00
                                                     − 00
                                                        0
```

Place the decimal point in the quotient.

2 **Divide 17.5 by 14.**

```
      .                      1.25
 14 )17.5    ──→       14 )17.50
                          − 14
                             3 5
                           − 2 8
                              70
                            − 70
                               0
```

Annex zeros in the dividend.

Divide until the remainder is 0.

Divide.

1. $9\overline{)12.6}$ **2.** $9\overline{)\$4.14}$ **3.** $4\overline{)\$23.64}$ **4.** $26\overline{)0.52}$

5. $16\overline{)25.6}$ **6.** $32\overline{)\$2.88}$ **7.** $9\overline{)27.54}$ **8.** $4\overline{)\$11.60}$

SKILL 15

Name ______________________ Date ____________ Period ________

Dividing Decimals by Whole Numbers *(continued)*

Divide.

9. $6\overline{)1.5}$

10. $18\overline{)25.2}$

11. $34\overline{)53.72}$

12. $14\overline{)37.8}$

13. $29\overline{)104.4}$

14. $34\overline{)12.92}$

15. $61\overline{)103.7}$

16. $74\overline{)26.64}$

17. $12\overline{)301.8}$

18. $33\overline{)89.1}$

19. $26\overline{)50.7}$

20. $15\overline{)\$62.40}$

21. $2.4 \div 96 =$

22. $5.59 \div 26 =$

23. $15.5 \div 50 =$

24. $34.55 \div 20 =$

25. $30.45 \div 35 =$

26. $27.93 \div 19 =$

27. $41.8 \div 55 =$

28. $411.84 \div 72 =$

Solve.

29. Eric bought an 8-ounce can of frozen orange juice on sale for \$0.72. What is the cost per ounce?

30. Lucy runs 4 miles in 22.7 minutes. What is her average time per mile?

SKILL 16

Name ______________________ Date ____________ Period ________

Dividing Decimals by Decimals

To divide by a decimal, change the divisor to a whole number.

Example **Find 0.5194 ÷ 0.49.**

$$\begin{array}{r} 1.06 \\ 0.49.\overline{)0.51.94} \\ \underline{49} \\ 2\,94 \\ \underline{2\,94} \\ 0 \end{array}$$

Change 0.49 to 49.
Move the decimal point two places to the right.
Move the decimal point in the dividend the same number of places to the right.
Divide as with whole numbers.

Without finding or changing each quotient, change each problem so that the divisor is a whole number.

1. 3.4 ÷ 1.1

2. 76.44 ÷ 0.006

3. 0.56 ÷ 0.4

4. 89.45 ÷ 0.908

5. 5.675 ÷ 6.8

6. 0.00864 ÷ 0.012

7. 0.84 ÷ 0.2

8. 1.02 ÷ 0.3

9. 3.9 ÷ 1.3

10. 13.6 ÷ 0.003

11. 1.622 ÷ 1.4

12. 0.00025 ÷ 0.035

Divide.

13. $0.9\overline{)6.3}$

14. $0.6\overline{)0.540}$

15. $0.3\overline{)129}$

16. $2.4\overline{)0.192}$

17. $0.44\overline{)5.28}$

18. $0.025\overline{)0.04}$

SKILL 16

Name ____________________ Date __________ Period ______

Dividing Decimals by Decimals *(continued)*

Divide.

19. $0.5\overline{)9.5}$ **20.** $0.8\overline{)0.048}$ **21.** $0.4\overline{)82}$

22. $3.5\overline{)2.38}$ **23.** $0.62\overline{)600.16}$ **24.** $0.015\overline{)0.06}$

25. $1.4\overline{)121.8}$ **26.** $8\overline{)0.0092}$ **27.** $0.38\overline{)760.38}$

28. $1.3\overline{)780}$ **29.** $0.08\overline{)0.0012}$ **30.** $0.7\overline{)5.95}$

Solve each equation.

31. $7.8 \div 2.6 = k$ **32.** $3.92 \div 0.08 = m$ **33.** $s = 149.73 \div 0.23$

34. $v = 155 \div 0.1$ **35.** $c = 1098 \div 6.1$ **36.** $3633.4 \div 3.7 = d$

37. $903.6 \div 25.1 = n$ **38.** $363.6 \div 5 = r$ **39.** $2.004 \div 0.2 = b$

40. $w = 84.7 \div 3.85$ **41.** $165.2 \div 8.26 = t$ **42.** $29.28 \div 1.22 = s$

43. y 5 0.0528 4 0.06 **44.** 16.84 4 0.4 5 m **45.** k 5 2.05 4 0.5

Name ______________________ Date __________ Period ______

Multiplying Decimals by Powers of 10

You can find the product of a decimal and a power of 10 without using a calculator or paper and pencil. Suppose you wanted to find the product of 36 and powers of 10.

Decimal		Power of Ten		Quotient
36	÷	10^{-3} or 0.001	=	0.036
36	÷	10^{-2} or 0.01	=	0.36
36	÷	10^{-1} or 0.1	=	3.6
36	÷	10^{0} or 1	=	36
36	÷	10^{1} or 10	=	360
36	÷	10^{2} or 100	=	3600
36	÷	10^{3} or 1000	=	36,000
36	÷	10^{4} or 10,000	=	360,000

For powers of 10 that are less than 1, the exponent in the power of 10 tells you the number of places to move the decimal point to the right. For powers of 10 that are greater than 1, the decimal point moves to the left.

Examples **1** $\mathbf{6 \cdot 10^3 = 6000}$ *Move the decimal point 3 places to the right.*

2 $\mathbf{4.5 \cdot 10^{-2} = 0.045}$ *Move the decimal point 2 places to the left.*

Multiply mentally.

1. $8 \cdot 0.01$

2. $55.8 \cdot 100$

3. $59 \cdot 10^4$

4. $14 \cdot 0.1$

5. $0.13 \cdot 10^{-3}$

6. $18 \cdot 10^2$

7. $17 \cdot 100$

8. $1.46 \cdot 0.001$

9. $12 \cdot 10^{-1}$

SKILL 17

Name ______________________ Date __________ Period ______

Multiplying Decimals by Powers of 10 *(continued)*

Multiply mentally.

10. $77 \cdot 1000$

11. $143 \cdot 100$

12. $15 \cdot 10$

13. $15 \cdot 10^0$

14. $1.36 \cdot 1000$

15. $184 \cdot 10^{-3}$

16. $1.7 \cdot 0.01$

17. $0.08 \cdot 10^{-2}$

18. $1432 \cdot 10^4$

19. $43 \cdot 10$

20. $13.5 \cdot 0.01$

21. $55 \cdot 10^{-2}$

22. $137 \cdot 100$

23. $43 \cdot 1000$

24. $281 \cdot 10^2$

Solve each equation.

25. $v = 78 \cdot 10$

26. $q = 654 \cdot 10^0$

27. $m = 198 \cdot 0.001$

28. $r = 876 \cdot 100$

29. $s = 15 \cdot 10^{-2}$

30. $t = 12.5 \cdot 0.01$

31. $p = 1.4 \cdot 1000$

32. $q = 385 \cdot 10^{-3}$

33. $u = 8.8 \cdot 10$

34. $14 \cdot 100 = r$

35. $w = 1.34 \cdot 10^3$

36. $k = 14.8 \cdot 0.1$

37. $n = 123 \cdot 0.1$

38. $4326 \cdot 10^0 = y$

39. $81.18 \cdot 10^{-3} = j$

40. $480 \cdot 10^4 = m$

41. $r = 6820 \cdot 10^1$

42. $q = 2.813 \cdot 10^{-2}$

Name ______________________ Date ____________ Period ________

Dividing Decimals by Powers of 10

You can find the quotient of a decimal and a power of 10 without using a calculator or paper and pencil. Suppose you wanted to find the quotient of 5540 and powers of 10.

Decimal		Power of Ten		Quotient
5540	÷	10^{-3} or 0.001	=	5,540,000
5540	÷	10^{-2} or 0.01	=	554,000
5540	÷	10^{-1} or 0.1	=	55,400
5540	÷	100 or 1	=	5540
5540	÷	10^1 or 10	=	554
5540	÷	10^2 or 100	=	55.4
5540	÷	10^3 or 1000	=	5.54
5540	÷	10^4 or 10,000	=	0.554

For powers of 10 that are less than 1, the exponent in the power of 10 tells you the number of places to move the decimal point to the left. For powers of 10 that are greater than 1, the decimal point moves to the right.

Examples **1** **8 ÷ 103 = 0.008** *Move the decimal point 3 places to the left.*

2 $\mathbf{0.34 \div 10^{-2} = 34}$ *Move the decimal point 2 places to the right.*

Divide mentally.

1. $6 \div 0.01$

2. $35.7 \div 100$

3. $764 \div 10^4$

4. $18 \div 0.1$

5. $0.145 \div 10^{-3}$

6. $24 \div 10^2$

7. $47 \div 100$

8. $1.53 \div 0.001$

9. $61 \div 10^{-1}$

SKILL 18

Name ______________________ Date ____________ Period ________

Dividing Decimals by Powers of 10 *(continued)*

Divide mentally.

10. $88 \div 1000$

11. $234 \div 100$

12. $34 \div 10$

13. $19 \div 10^0$

14. $1.27 \div 1000$

15. $765 \div 10^{-3}$

16. $1.1 \div 0.01$

17. $0.04 \div 10^{-2}$

18. $1561 \div 10^4$

19. $54 \div 10$

20. $15.2 \div 0.01$

21. $66 \div 10^{-2}$

22. $128 \div 100$

23. $55{,}510 \div 1000$

24. $426 \div 10^2$

Solve each equation.

25. $v = 87 \div 10$

26. $q = 737 \div 10^0$

27. $m = 891 \div 0.001$

28. $r = 678 \div 100$

29. $s = 24 \div 10^{-2}$

30. $t = 16.4 \div 0.01$

31. $p = 1.3 \div 1000$

32. $q = 0.573 \div 10^{-3}$

33. $u = 9.9 \div 10$

34. $148 \div 100 = r$

35. $w = 1.28 \div 10^3$

36. $k = 16.5 \div 0.1$

37. $n = 154 \div 0.1$

38. $3546 \div 10^0 = y$

39. $41.14 \div 10^{-3} = j$

40. $360 \div 10^4 = m$

41. $r = 7610 \div 10^1$

42. $q = 2.532 \div 10^{-2}$

SKILL 19

Name ____________ Date ________ Period ______

Equivalent Fractions

To find equivalent fractions, multiply or divide the numerator and denominator by the same nonzero number.

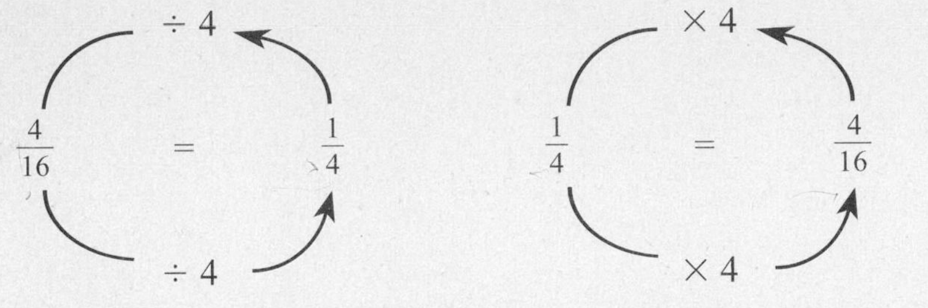

The shaded region at the right shows that $\frac{4}{16}$ and $\frac{1}{4}$ are equivalent.

Examples

1 Complete $\frac{9}{12} = \frac{18}{}$ so that the fractions are equivalent.

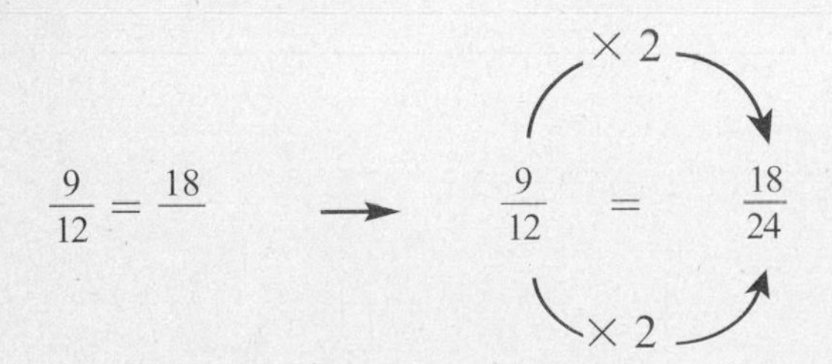

Since 9 × 2 = 18, multiply both the numerator and the denominator by 2.

2 Find three fractions equivalent to $\frac{4}{9}$.

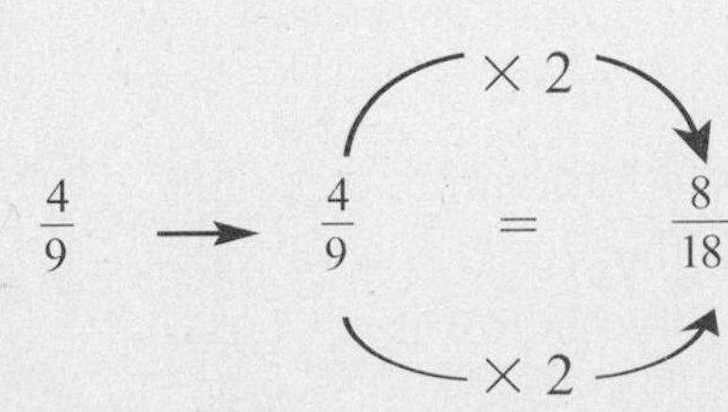

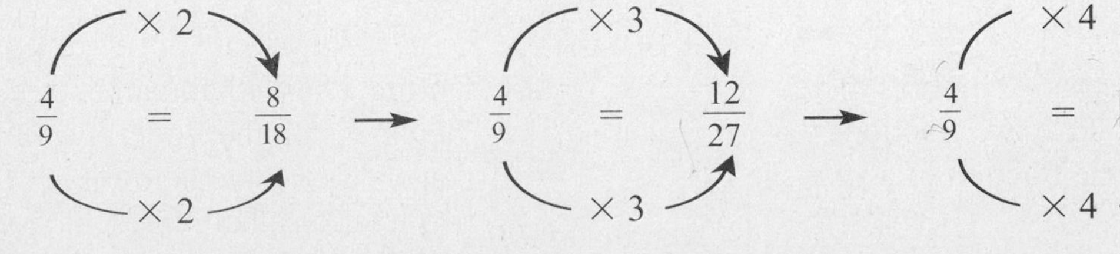

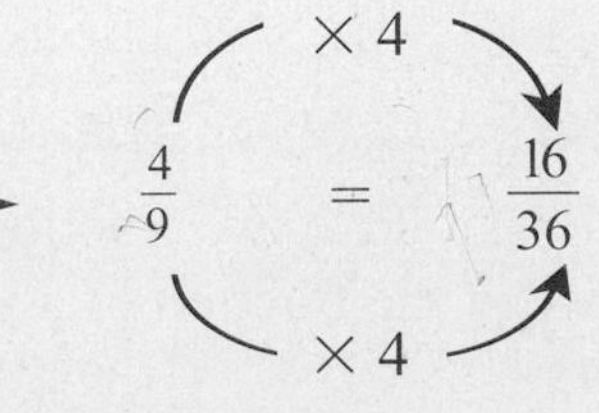

Complete so that the fractions are equivalent.

1. $\frac{3}{4} = \frac{}{12}$ **2.** $\frac{4}{9} = \frac{}{18}$ **3.** $\frac{4}{5} = \frac{}{20}$ **4.** $\frac{5}{8} = \frac{}{24}$

SKILL 19

Name ______________________ Date ____________ Period ________

Equivalent Fractions *(continued)*

Complete so that the fractions are equivalent.

5. $\frac{3}{5} = \frac{15}{}$ **6.** $\frac{5}{7} = \frac{10}{}$ **7.** $\frac{4}{9} = \frac{12}{}$ **8.** $\frac{3}{8} = \frac{6}{}$

9. $\frac{2}{3} = \frac{}{24}$ **10.** $\frac{5}{15} = \frac{}{3}$ **11.** $\frac{5}{20} = \frac{}{4}$ **12.** $\frac{7}{56} = \frac{}{8}$

13. $\frac{16}{40} = \frac{2}{}$ **14.** $\frac{27}{72} = \frac{3}{}$ **15.** $\frac{40}{64} = \frac{5}{}$ **16.** $\frac{10}{45} = \frac{2}{}$

17. $\frac{16}{18} = \frac{8}{}$ **18.** $\frac{4}{7} = \frac{}{42}$ **19.** $\frac{6}{11} = \frac{}{33}$ **20.** $\frac{5}{12} = \frac{25}{}$

Find three fractions equivalent to each of the following.

21. $\frac{1}{2}$ **22.** $\frac{4}{5}$

23. $\frac{2}{3}$ **24.** $\frac{5}{6}$

25. $\frac{7}{8}$ **26.** $\frac{9}{10}$

Solve.

27. Ms. Yen works 10 months of 12 each year. Give two fractions that represent the fraction of a year she works.

28. During a basketball game, there are 10 players on the floor. Five of the players are on the home team. Give two fractions that represent the fraction of players on the floor that are on the home team.

Name ______________________ Date ____________ Period ________

Simplifying Fractions

To write a fraction in simplest form, divide both the numerator and denominator by their greatest common factor (GCF).

Example 1 **Write $\frac{16}{100}$ in simplest form.**

Step 1	Step 2
Find the GCF of 16 and 100. You can use prime factorization. $16 = 2 \times 2 \times 2 \times 2$ $100 = 2 \times 2 \times 5 \times 5$ GCF: $2 \times 2 = 4$	Divide both 16 and 100 by their GCF, 4. $\frac{16}{100} = \frac{4}{25}$ (÷ 4 in numerator and denominator) *A fraction is in simplest form when the GCF of both its numerator and denominator is 1.*

The fraction $\frac{16}{100}$ written in simplest from is $\frac{4}{25}$.

Example 2 **Write $\frac{6}{15}$ in simplest form.**

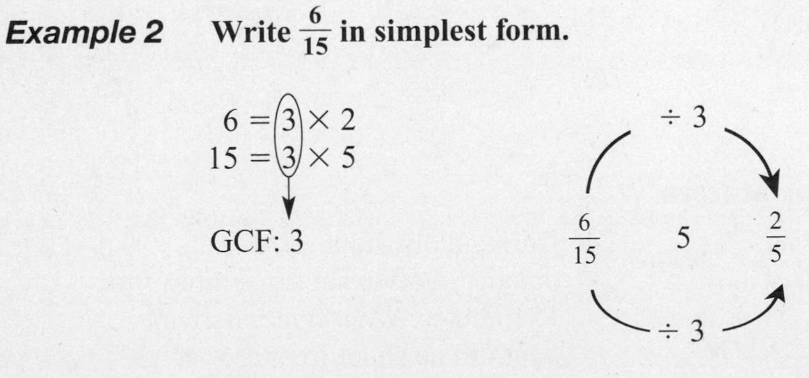

Write each fraction in simplest form.

1. $\frac{4}{6}$
2. $\frac{2}{4}$
3. $\frac{6}{12}$
4. $\frac{8}{10}$
5. $\frac{6}{14}$
6. $\frac{6}{9}$
7. $\frac{2}{8}$
8. $\frac{3}{12}$

SKILL 20

Name ______________ Date ________ Period ______

Simplifying Fractions *(continued)*

Write each fraction in simplest form.

9. $\frac{13}{26}$ **10.** $\frac{16}{24}$ **11.** $\frac{12}{18}$ **12.** $\frac{12}{16}$

13. $\frac{5}{15}$ **14.** $\frac{15}{25}$ **15.** $\frac{3}{15}$ **16.** $\frac{10}{30}$

17. $\frac{9}{21}$ **18.** $\frac{14}{30}$ **19.** $\frac{20}{36}$ **20.** $\frac{6}{24}$

21. $\frac{27}{9}$ **22.** $\frac{10}{100}$ **23.** $\frac{25}{40}$ **24.** $\frac{8}{16}$

25. $\frac{10}{25}$ **26.** $\frac{8}{40}$ **27.** $\frac{12}{30}$ **28.** $\frac{16}{20}$

29. $\frac{7}{42}$ **30.** $\frac{15}{30}$ **31.** $\frac{9}{33}$ **32.** $\frac{10}{16}$

Solve. Write the answer in simplest form.

33. Tara takes 12 vacation days in June, which has 30 days. What fraction of the month is she on vacation? Express your answer in simplest form.

34. During a one-hour (60 minute) practice, Calvin shot free throws for 15 minutes. What fraction of an hour did he shoot free throws? Express your answer in simplest form.

Name ____________ Date ________ Period ______

Writing Improper Fractions as Mixed Numbers

A fraction such as $\frac{8}{5}$ is called an **improper fraction** because the numerator is greater than the denominator. Improper fractions are often expressed as mixed numbers. A **mixed number** is the sum of a whole number and a fraction. Follow the steps in Example 1 to write $\frac{8}{5}$ as a mixed number.

Example 1 **Write $\frac{8}{5}$ as a mixed number in simplest form.**

Step 1	Step 2
Divide the numerator by the denominator. $\begin{array}{r} 1 \\ 5\overline{)8} \\ -5 \\ \hline 3 \end{array}$	Write the remainder as a fraction. $\begin{array}{r} 1\frac{3}{5} \\ 5\overline{)8} \\ -5 \\ \hline 3 \end{array}$

Example 2 **Write $\frac{38}{4}$ as a mixed number in simplest form.**

$$\begin{array}{r} 9\frac{2}{4} = 9\frac{1}{2} \\ 4\overline{)38} \\ -36 \\ \hline 2 \end{array}$$

Write each improper fraction as a mixed number in simplest form.

1. $\frac{7}{5}$

2. $\frac{13}{8}$

3. $\frac{13}{4}$

4. $\frac{22}{7}$

5. $\frac{6}{4}$

6. $\frac{14}{8}$

7. $\frac{9}{6}$

8. $\frac{14}{10}$

SKILL 21

Name ____________________ Date __________ Period ______

Writing Improper Fractions as Mixed Numbers *(continued)*

Write each improper fraction as a mixed number in simplest form.

9. $\frac{28}{16}$ **10.** $\frac{25}{10}$ **11.** $\frac{33}{9}$ **12.** $\frac{40}{16}$

13. $\frac{13}{5}$ **14.** $\frac{9}{2}$ **15.** $\frac{15}{3}$ **16.** $\frac{21}{8}$

17. $\frac{17}{12}$ **18.** $\frac{12}{5}$ **19.** $\frac{13}{3}$ **20.** $\frac{15}{10}$

21. $\frac{28}{12}$ **22.** $\frac{21}{5}$ **23.** $\frac{19}{6}$ **24.** $\frac{31}{8}$

25. $\frac{16}{5}$ **26.** $\frac{27}{15}$ **27.** $\frac{32}{12}$ **28.** $\frac{48}{24}$

29. $\frac{36}{24}$ **30.** $\frac{25}{20}$ **31.** $\frac{30}{12}$ **32.** $\frac{24}{10}$

Solve. Write each answer as a mixed number in simplest form.

33. Carrie rode her bike 22 miles in 3 hours. What is the average number of miles she rode in one hour?

34. Mr. Steele has managed the Classic Theater for 21 months. How many years has he managed the Classic Theater?

Name ______________________ Date ____________ Period ________

Writing Mixed Numbers as Improper Fractions

Follow the steps in Example 1 to change a mixed number to an improper fraction.

Example 1 **Write $3\frac{1}{2}$ as an improper fraction.**

Step 1	Step 2
First multiply the whole number by the denominator and add the numerator. Then write this sum over the denominator. $3\frac{1}{2} = \frac{(3 \times 2) + 1}{2}$	Simplify. $\frac{(3 \times 2) + 1}{2} = \frac{6 + 1}{2}$ or $\frac{7}{2}$

Example 2 **Write $8\frac{3}{5}$ as an improper fraction.**

$$8\frac{3}{5} = \frac{(5 \times 8) + 3}{5} = \frac{43}{5}$$

Write each mixed number as an improper fraction.

1. $6\frac{1}{3}$ **2.** $5\frac{3}{4}$ **3.** $7\frac{1}{6}$ **4.** $9\frac{1}{8}$

5. $2\frac{3}{16}$ **6.** $4\frac{3}{10}$ **7.** $4\frac{2}{3}$ **8.** $3\frac{3}{5}$

9. $5\frac{6}{7}$ **10.** $3\frac{7}{9}$ **11.** $2\frac{11}{12}$ **12.** $4\frac{7}{8}$

SKILL 22

Name ______________________ Date ____________ Period ________

Writing Mixed Numbers as Improper Fractions *(continued)*

Write each mixed number as an improper fraction.

13. $1\frac{3}{8}$ **14.** $5\frac{2}{5}$ **15.** $2\frac{3}{4}$ **16.** $1\frac{7}{8}$

17. $1\frac{7}{12}$ **18.** $4\frac{1}{2}$ **19.** $2\frac{9}{10}$ **20.** $3\frac{5}{8}$

21. $3\frac{2}{3}$ **22.** $4\frac{3}{4}$ **23.** $5\frac{2}{3}$ **24.** $5\frac{1}{8}$

25. $5\frac{9}{10}$ **26.** $6\frac{7}{8}$ **27.** $4\frac{3}{10}$ **28.** $10\frac{2}{3}$

29. $9\frac{7}{12}$ **30.** $8\frac{5}{11}$ **31.** $15\frac{2}{7}$ **32.** $12\frac{4}{7}$

33. $11\frac{4}{5}$ **34.** $18\frac{2}{3}$ **35.** $20\frac{1}{4}$ **36.** $16\frac{4}{9}$

37. $5\frac{12}{13}$ **38.** $16\frac{2}{13}$ **39.** $24\frac{1}{3}$ **40.** $8\frac{16}{17}$

41. $9\frac{5}{17}$ **42.** $7\frac{6}{19}$ **43.** $5\frac{8}{9}$ **44.** $16\frac{10}{13}$

Name ________________ Date ________ Period ______

Comparing and Ordering Fractions

One way to compare fractions is to express them as fractions with the same denominator. The **least common denominator (LCD)** is the least common multiple of the denominators.

Example **Replace the ◯ with $<$, $>$, or $=$ to make a true sentence.**

$\frac{5}{8}$ ◯ $\frac{2}{3}$

The LCM of 8 and 3 is 24. Express $\frac{5}{8}$ and $\frac{2}{3}$ as fractions with a denominator of 24.

$\frac{15}{24}$ ◯ $\frac{16}{24}$ Compare the numerators. Since $15 < 16$, $\frac{15}{24} < \frac{16}{24}$. Therefore, $\frac{5}{8} < \frac{2}{3}$.

Find the LCD for each pair of fractions.

1. $\frac{2}{5}, \frac{1}{3}$
2. $\frac{3}{4}, \frac{5}{6}$
3. $\frac{1}{2}, \frac{4}{7}$
4. $\frac{4}{5}, \frac{2}{3}$
5. $\frac{5}{8}, \frac{7}{12}$
6. $\frac{1}{2}, \frac{6}{7}$
7. $\frac{1}{6}, \frac{9}{10}$
8. $\frac{3}{4}, \frac{2}{9}$
9. $\frac{5}{12}, \frac{3}{16}$

Replace each ◯ with $<$, $>$, or $=$ to make a true sentence.

10. $\frac{3}{4}$ ◯ $\frac{4}{5}$
11. $\frac{3}{8}$ ◯ $\frac{9}{24}$
12. $\frac{2}{3}$ ◯ $\frac{9}{15}$
13. $\frac{7}{12}$ ◯ $\frac{2}{3}$
14. $\frac{5}{11}$ ◯ $\frac{1}{3}$
15. $\frac{27}{36}$ ◯ $\frac{3}{4}$

SKILL 23

Name ______________________ Date __________ Period ______

Comparing and Ordering Fractions *(continued)*

Replace each ◯ with <, >, or = to make a true sentence.

16. $\frac{5}{6} \bigcirc \frac{7}{8}$

17. $\frac{6}{7} \bigcirc \frac{4}{5}$

18. $\frac{3}{9} \bigcirc \frac{1}{3}$

19. $\frac{5}{8} \bigcirc \frac{7}{12}$

20. $\frac{5}{7} \bigcirc \frac{7}{10}$

21. $\frac{2}{3} \bigcirc \frac{3}{4}$

22. $\frac{2}{15} \bigcirc \frac{1}{6}$

23. $\frac{3}{8} \bigcirc \frac{6}{16}$

24. $\frac{5}{12} \bigcirc \frac{2}{5}$

25. $\frac{3}{10} \bigcirc \frac{5}{14}$

26. $\frac{4}{9} \bigcirc \frac{3}{7}$

27. $\frac{1}{6} \bigcirc \frac{2}{12}$

28. $\frac{3}{5} \bigcirc \frac{5}{9}$

29. $\frac{7}{9} \bigcirc \frac{4}{7}$

30. $\frac{9}{10} \bigcirc \frac{11}{12}$

31. $\frac{1}{4} \bigcirc \frac{2}{8}$

32. $\frac{2}{9} \bigcirc \frac{4}{15}$

33. $\frac{8}{9} \bigcirc \frac{7}{8}$

Order the following fractions from least to greatest.

34. $\frac{3}{4}, \frac{2}{5}, \frac{5}{8}, \frac{1}{2}$

35. $\frac{2}{3}, \frac{4}{9}, \frac{5}{6}, \frac{7}{12}$

36. $\frac{1}{3}, \frac{2}{7}, \frac{3}{14}, \frac{1}{6}$

37. $\frac{7}{15}, \frac{3}{5}, \frac{5}{12}, \frac{1}{2}$

38. $\frac{11}{12}, \frac{5}{6}, \frac{3}{4}, \frac{9}{16}$

39. $\frac{4}{5}, \frac{2}{3}, \frac{11}{35}, \frac{7}{9}$

40. $\frac{7}{8}, \frac{4}{5}, \frac{3}{4}, \frac{9}{10}$

41. $\frac{1}{3}, \frac{2}{5}, \frac{3}{12}, \frac{3}{10}$

42. $\frac{1}{2}, \frac{3}{5}, \frac{2}{7}, \frac{5}{9}$

43. $\frac{1}{10}, \frac{2}{3}, \frac{1}{12}, \frac{5}{6}$

Name ______________________ Date ____________ Period ________

Multiplying Fractions

To multiply fractions, multiply the numerators. Then multiply the denominators. Simplify the product if possible.

Examples **1** **Multiply $\frac{4}{7}$ times $\frac{5}{9}$.**

$\frac{4}{7} \times \frac{5}{9} = \frac{4 \times 5}{7 \times 9}$ *Multiply the numerators. Multiply the denominators.*

$= \frac{20}{63}$

The product of $\frac{4}{7}$ and $\frac{5}{9}$ is $\frac{20}{63}$.

2 **Multiply $\frac{5}{6}$ times $\frac{3}{5}$.**

$\frac{5}{6} \times \frac{3}{5} = \frac{5 \times 3}{6 \times 5}$ *Multiply the numerators. Multiply the denominators.*

$= \frac{15}{30}$ or $\frac{1}{2}$ *Simplify.*

The product of $\frac{5}{6}$ and $\frac{3}{5}$ is $\frac{1}{2}$.

Multiply.

1. $\frac{2}{3} \times \frac{1}{4}$ **2.** $\frac{3}{7} \times \frac{1}{2}$ **3.** $\frac{1}{3} \times \frac{3}{5}$

4. $\frac{1}{2} \times \frac{6}{7}$ **5.** $\frac{7}{10} \times \frac{5}{7}$ **6.** $\frac{1}{4} \times \frac{1}{4}$

7. $\frac{1}{3} \times \frac{1}{5}$ **8.** $\frac{5}{8} \times \frac{1}{2}$ **9.** $\frac{4}{9} \times \frac{3}{4}$

10. $\frac{2}{3} \times \frac{3}{8}$ **11.** $\frac{1}{7} \times \frac{1}{7}$ **12.** $\frac{2}{9} \times \frac{1}{2}$

13. $\frac{3}{5} \times \frac{5}{6}$ **14.** $\frac{2}{7} \times \frac{1}{3}$ **15.** $\frac{5}{12} \times \frac{1}{5}$

16. $\frac{1}{2} \times \frac{1}{5}$ **17.** $\frac{6}{7} \times \frac{8}{15}$ **18.** $\frac{8}{9} \times \frac{9}{10}$

19. $\frac{4}{5} \times \frac{5}{14}$ **20.** $\frac{7}{8} \times \frac{4}{9}$ **21.** $\frac{5}{8} \times \frac{3}{4}$

Name ______________________ Date ____________ Period ______

Multiplying Fractions *(continued)*

Use the recipe for lemon chicken saute below to answer Exercises 22–25.

6 boneless chicken breasts, rolled in flour	$\frac{1}{3}$ cup teriyaki sauce
$\frac{1}{4}$ cup butter	$\frac{1}{2}$ teaspoon sugar
3 tablespoons lemon juice	$\frac{1}{8}$ teaspoon pepper
1 teaspoon garlic	

22. If Julie wants to make half of this recipe, how much pepper should she use?

23. If Julie wants to make one-third of this recipe, how much teriyaki sauce should she use?

24. If Julie wants to make two-thirds of this recipe, how much sugar should she use?

25. If Julie wants to make two-thirds of this recipe, how much butter should she use?

26. If about $\frac{1}{3}$ of Earth is able to be farmed and $\frac{2}{5}$ of this land is planted in grain crops, what part of Earth is planted in grain crops?

27. Two fifths of the students at Main Street Middle School are in seventh grade. If half of the students in seventh grade are boys, what fraction of the students are seventh grade boys?

SKILL 25

Name ____________ Date ________ Period ______

Multiplying Fractions and Mixed Numbers

To multiply fractions: Multiply the numerators. Then multiply the denominators.

$$\frac{5}{6} \times \frac{3}{5} = \frac{5 \times 3}{6 \times 5} = \frac{15}{30} = \frac{1}{2}$$

To multiply mixed numbers: Rename each mixed number as a fraction. Multiply the fractions.

$$7 \times 1\frac{1}{4} = \frac{7}{1} \times \frac{5}{4} = \frac{35}{4} = 8\frac{3}{4}$$

Multiply. Write each product in simplest form.

1. $\frac{2}{3} \times \frac{1}{4}$

2. $\frac{3}{7} \times \frac{1}{2}$

3. $\frac{1}{3} \times \frac{3}{5}$

4. $\frac{1}{2} \times \frac{6}{7}$

5. $\frac{3}{8} \times 4$

6. $\frac{7}{10} \times \frac{5}{7}$

7. $\frac{4}{9} \times 3$

8. $\frac{1}{4} \times \frac{1}{4}$

9. $1\frac{1}{2} \times 6$

10. $\frac{3}{4} \times 1\frac{2}{3}$

11. $3\frac{1}{3} \times 2\frac{1}{2}$

12. $4\frac{1}{5} \times \frac{1}{7}$

SKILL 25

Name ______________________ Date ____________ Period ______

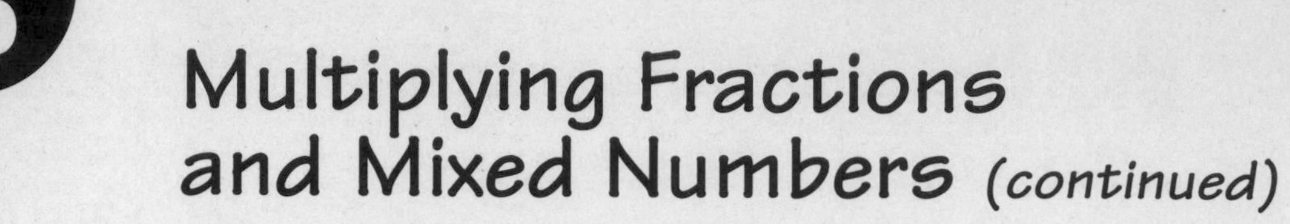

Multiplying Fractions and Mixed Numbers *(continued)*

Multiply. Write each product in simplest form.

13. $1\frac{1}{9} \times \frac{3}{5}$

14. $6 \times \frac{11}{12}$

15. $\frac{1}{2} \times 2\frac{2}{3}$

16. $\frac{2}{3} \times \frac{1}{2}$

17. $\frac{3}{4} \times \frac{1}{9}$

18. $3 \times \frac{4}{9}$

19. $\frac{1}{5} \times \frac{1}{4}$

20. $\frac{1}{4} \times \frac{4}{5}$

21. $\frac{4}{9} \times \frac{3}{4}$

22. $\frac{13}{21} \times \frac{7}{13}$

23. $\frac{7}{8} \times \frac{4}{9}$

24. $\frac{5}{7} \times \frac{7}{10}$

25. $\frac{4}{5} \times \frac{5}{14}$

26. $\frac{1}{4} \times \frac{5}{8}$

27. $\frac{2}{3} \times \frac{5}{9}$

28. $\frac{4}{5} \times 7$

29. $2\frac{2}{5} \times 1\frac{3}{7}$

30. $6 \times \frac{2}{3}$

31. $3\frac{3}{4} \times \frac{1}{2}$

32. $1\frac{5}{9} \times 2\frac{4}{7}$

33. $4\frac{1}{3} \times \frac{1}{2}$

Name ______________________ Date __________ Period ______

Dividing Fractions

To divide by a fraction, multiply by its reciprocal.
Simplify the quotient if possible.

Examples ***1*** **Divide $\frac{2}{3}$ by $\frac{5}{7}$.**

$\frac{2}{3} \div \frac{5}{7} = \frac{2}{3} \times \frac{7}{5}$ *Multiply by the reciprocal of* $\frac{5}{7}$.

$= \frac{2 \times 7}{3 \times 5}$ *Multiply the numerators.*
Multiply the denominators.

$= \frac{14}{15}$

The quotient is $\frac{14}{15}$.

2 **Divide $\frac{3}{4}$ by $\frac{9}{10}$.**

$\frac{3}{4} \div \frac{9}{10} = \frac{3}{4} \times \frac{10}{9}$ *Multiply by the reciprocal of* $\frac{9}{10}$.

$= \frac{3 \times 10}{4 \times 9}$ *Multiply the numerators.*
Multiply the denominators.

$= \frac{30}{36}$ or $\frac{5}{6}$ *Simplify.*

The quotient is $\frac{5}{6}$.

Divide.

1. $\frac{3}{4} \div \frac{1}{2}$

2. $\frac{4}{5} \div \frac{1}{3}$

3. $\frac{1}{5} \div \frac{1}{4}$

4. $\frac{4}{7} \div \frac{8}{9}$

5. $\frac{3}{8} \div \frac{3}{4}$

6. $\frac{9}{7} \div \frac{3}{14}$

7. $\frac{4}{5} \div \frac{2}{5}$

8. $\frac{7}{8} \div \frac{1}{4}$

9. $\frac{2}{5} \div \frac{5}{8}$

Name ____________ Date ________ Period ______

Dividing Fractions *(continued)*

Divide.

10. $\frac{1}{3} \div \frac{1}{6}$

11. $\frac{5}{8} \div \frac{5}{12}$

12. $\frac{4}{5} \div \frac{2}{7}$

13. $\frac{2}{5} \div \frac{3}{10}$

14. $\frac{5}{7} \div \frac{3}{4}$

15. $\frac{2}{3} \div \frac{4}{9}$

16. $\frac{4}{7} \div \frac{4}{5}$

17. $\frac{5}{6} \div \frac{1}{9}$

18. $\frac{4}{5} \div \frac{2}{3}$

19. About $\frac{1}{20}$ of the population of the world lives in South America. If about $\frac{1}{35}$ of the population of the world lives in Brazil, what fraction of the population of South America lives in Brazil?

20. Three fourths of a pizza is left. If the pizza was originally cut in $\frac{1}{8}$ pieces, how many pieces are left?

The area of each rectangle is given. Find the missing length for each rectangle.

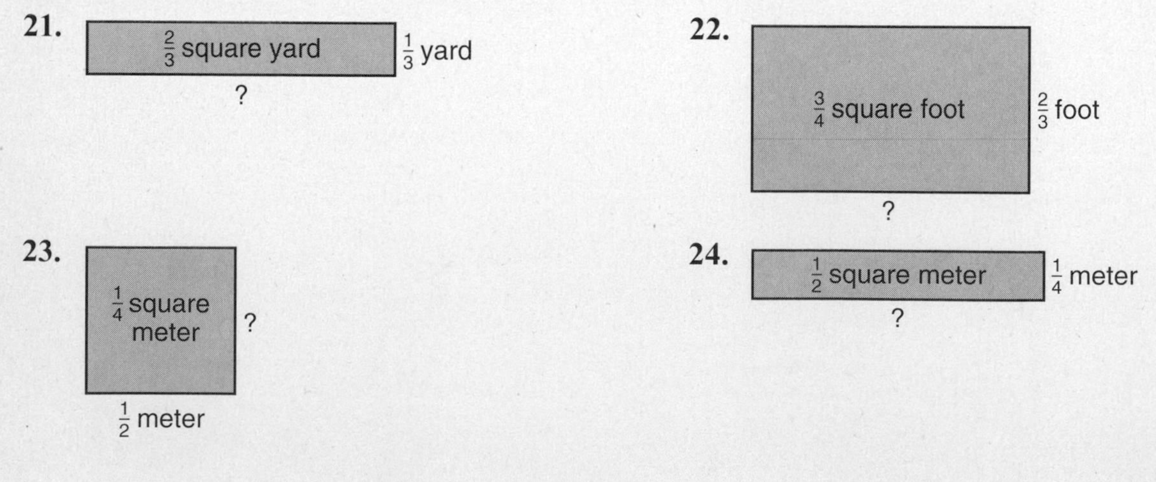

Name ______________________ Date __________ Period ______

Dividing Fractions and Mixed Numbers

To divide fractions and mixed numbers:

1. Write any mixed numbers as improper fractions.
2. Find the reciprocal of the divisor.
3. Multiply the dividend by the reciprocal of the divisor.

Examples **1** $\mathbf{\frac{5}{8} \div \frac{5}{12}}$ *The reciprocal of* $\frac{5}{12}$ *is* $\frac{12}{5}$.

$$\frac{5}{8} \div \frac{5}{12} = \frac{5}{8} \times \frac{12}{5}$$
$$= \frac{60}{40} \text{ or } 1\frac{1}{2}$$

2 $\mathbf{7 \div 3\frac{1}{2} \longrightarrow \frac{7}{1} \div \frac{7}{2}}$ *The reciprocal of* $\frac{7}{2}$ *is* $\frac{2}{7}$.

$$7 \div 3\frac{1}{2} = \frac{7}{1} \times \frac{2}{7}$$
$$= \frac{14}{7} \text{ or } 2$$

Name the reciprocal of each number.

1. $\frac{6}{11}$ **2.** $\frac{14}{5}$ **3.** 8 **4.** $\frac{1}{5}$

Divide. Write each quotient in simplest form.

5. $\frac{7}{8} \div \frac{1}{4}$ **6.** $\frac{2}{5} \div \frac{5}{8}$ **7.** $\frac{1}{3} \div \frac{1}{6}$

8. $8 \div \frac{1}{3}$ **9.** $\frac{5}{9} \div 5$ **10.** $\frac{2}{4} \div 1\frac{1}{2}$

11. $2\frac{1}{2} \div 5$ **12.** $3\frac{1}{3} \div \frac{2}{9}$ **13.** $\frac{5}{8} \div 2\frac{1}{2}$

SKILL 27

Name ______________________ Date ____________ Period ________

Dividing Fractions and Mixed Numbers *(continued)*

Divide. Write each quotient in simplest form.

14. $1\frac{1}{3} \div 2\frac{1}{2}$

15. $3\frac{1}{3} \div 1\frac{2}{5}$

16. $\frac{9}{10} \div 5\frac{2}{5}$

17. $\frac{7}{8} \div \frac{2}{3}$

18. $5 \div \frac{3}{5}$

19. $3\frac{1}{4} \div 2\frac{1}{3}$

Solve each equation. Write each answer in simplest form.

20. $s = \frac{3}{4} \div \frac{1}{2}$

21. $k = \frac{4}{5} \div \frac{1}{3}$

22. $\frac{1}{5} \div \frac{1}{4} = y$

23. $u = 4 \div \frac{1}{3}$

24. $\frac{4}{7} \div \frac{8}{9} = j$

25. $w = \frac{3}{8} \div \frac{3}{4}$

26. $\frac{9}{7} \div 1\frac{3}{4} = h$

27. $\frac{4}{5} \div \frac{2}{5} = p$

28. $5 \div 3\frac{3}{4} = q$

29. $c = \frac{3}{8} \div 2\frac{1}{4}$

30 $t = 7\frac{1}{3} \div 4$

31. $m = 3\frac{1}{4} \div 2\frac{1}{4}$

32. $n = 1\frac{2}{7} \div 1\frac{13}{14}$

33. $1\frac{1}{5} \div \frac{3}{10} = r$

34. $7\frac{1}{2} \div 2\frac{5}{6} = w$

SKILL 28

Name ______________________ Date __________ Period ______

Adding Fractions

To add fractions with like denominators, add the numerators. Write the sum over the common denominator. Simplify the sum if possible.

Example 1 **Add: $\frac{7}{8} + \frac{5}{8}$.**

$$\begin{array}{r} \frac{7}{8} \\ + \frac{5}{8} \\ \hline \frac{12}{8} = \frac{3}{2} \text{ or } 1\frac{1}{2} \end{array}$$

Simplify the sum.

To add fractions with unlike denominators, rename the fractions with a common denominator. Then add the fractions.

Example 2 **Add: $\frac{1}{9} + \frac{5}{6}$.**

$$\begin{array}{rcl} \frac{1}{9} & = & \frac{2}{18} \\ + \frac{5}{6} & = & \frac{15}{18} \\ \hline & & \frac{17}{18} \end{array}$$

Use 18 for the common denominator.

Add.

1. $\begin{array}{r} \frac{4}{7} \\ + \frac{2}{7} \\ \hline \end{array}$

2. $\begin{array}{r} \frac{5}{9} \\ + \frac{4}{9} \\ \hline \end{array}$

3. $\begin{array}{r} \frac{11}{15} \\ + \frac{2}{15} \\ \hline \end{array}$

4. $\begin{array}{r} \frac{11}{15} \\ + \frac{7}{15} \\ \hline \end{array}$

5. $\begin{array}{r} \frac{6}{7} \\ + \frac{6}{7} \\ \hline \end{array}$

6. $\begin{array}{r} \frac{11}{12} \\ + \frac{5}{12} \\ \hline \end{array}$

Name ____________________ Date __________ Period ______

Adding Fractions *(continued)*

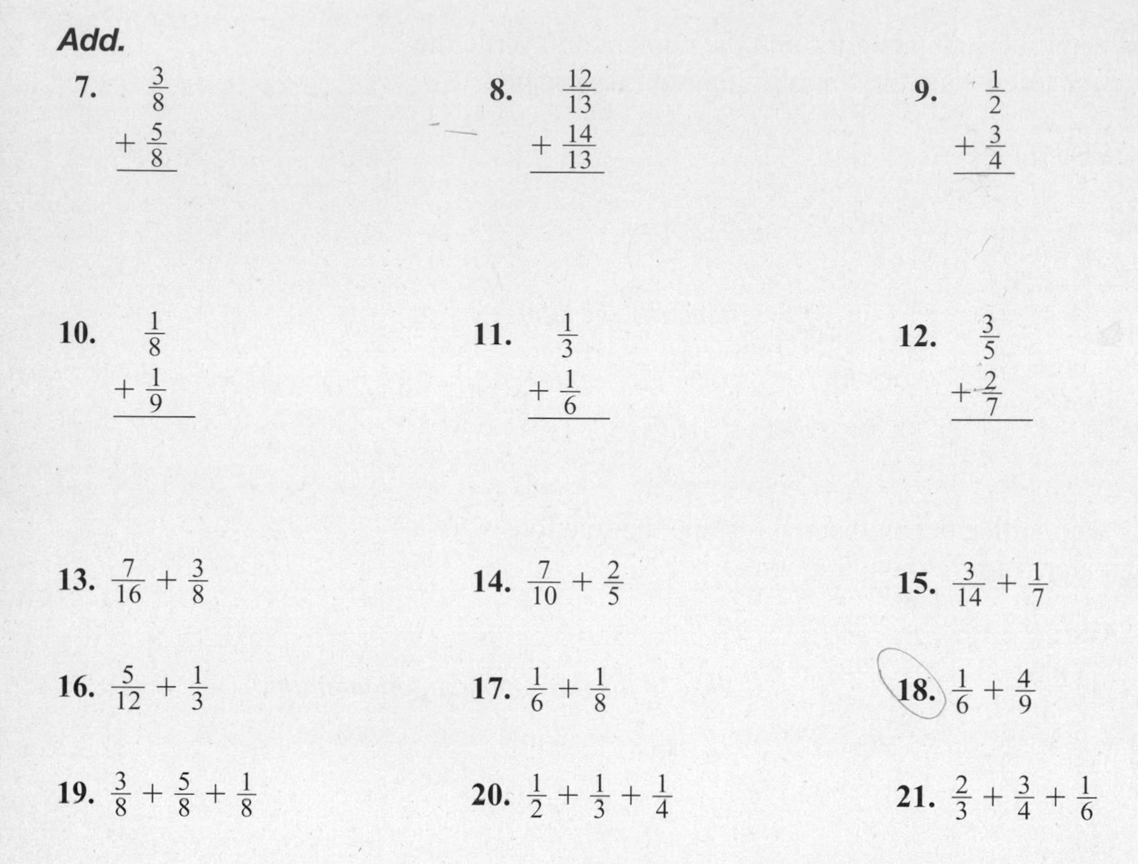

Add.

7. $\frac{3}{8} + \frac{5}{8}$

8. $\frac{12}{13} + \frac{14}{13}$

9. $\frac{1}{2} + \frac{3}{4}$

10. $\frac{1}{8} + \frac{1}{9}$

11. $\frac{1}{3} + \frac{1}{6}$

12. $\frac{3}{5} + \frac{2}{7}$

13. $\frac{7}{16} + \frac{3}{8}$

14. $\frac{7}{10} + \frac{2}{5}$

15. $\frac{3}{14} + \frac{1}{7}$

16. $\frac{5}{12} + \frac{1}{3}$

17. $\frac{1}{6} + \frac{1}{8}$

18. $\frac{1}{6} + \frac{4}{9}$

19. $\frac{3}{8} + \frac{5}{8} + \frac{1}{8}$

20. $\frac{1}{2} + \frac{1}{3} + \frac{1}{4}$

21. $\frac{2}{3} + \frac{3}{4} + \frac{1}{6}$

22. After running $\frac{7}{8}$ mile in a horse race, a horse ran an additional $\frac{3}{8}$ mile to cool down. How far did the horse run altogether?

23. In 1991, about $\frac{1}{5}$ of the crude oil produced was from North America, and about $\frac{2}{7}$ of the crude oil produced was from the Middle East.
What fraction of the crude oil produced was from North America or the Middle East?

24. In 1991, about $\frac{3}{10}$ of the petroleum consumed was in North America, and about $\frac{1}{5}$ of the petroleum consumed was in Western Europe. What fraction of the petroleum consumed was in North America or Western Europe?

Name ______________________ Date __________ Period ______

Adding Fractions and Mixed Numbers

To add fractions and mixed numbers, first rename each fraction as necessary. Then add the fractions. Next, add the whole numbers. Rename and simplify if necessary.

Example 1 **Add: $4\frac{5}{6} + 5\frac{1}{4}$.**

Step 1	Step 2	Step 3
Rename each fraction by finding the LCD if necessary. $4\frac{5}{6} \rightarrow 4\frac{10}{12}$; $+5\frac{1}{4} \rightarrow +5\frac{3}{12}$	Add the fractions. Then add the whole numbers. $4\frac{10}{12} + 5\frac{3}{12} = 9\frac{13}{12}$	Rename and simplify if necessary. $9\frac{13}{12} = 10\frac{1}{12}$

Example 2 **Add: $14\frac{5}{9} + 7$.**

$$\begin{array}{r} 14\frac{5}{9} \\ +\ 7 \\ \hline 21\frac{5}{9} \end{array}$$

Add. Write each sum in simplest form.

1. $13 + 9\frac{7}{8}$

2. $6\frac{1}{4} + 8\frac{3}{4}$

3. $5\frac{1}{6} + 7\frac{1}{3}$

4. $11\frac{3}{4} + 8\frac{2}{3}$

5. $16\frac{1}{2} + 14\frac{5}{7}$

6. $15\frac{1}{2} + 9\frac{4}{5}$

7. $18\frac{7}{8} + 15\frac{5}{8}$

8. $12\frac{1}{10} + 7\frac{5}{6}$

SKILL 29

Name ______________ Date ________ Period ______

Adding Fractions and Mixed Numbers *(continued)*

Add. Write each sum in simplest form.

9. $18\frac{7}{8} + 13$

10. $11 + 3\frac{5}{9}$

11. $9\frac{7}{9} + 3\frac{4}{9}$

12. $8\frac{2}{5} + 4\frac{4}{5}$

13. $12\frac{1}{2} + 8\frac{2}{3}$

14. $14\frac{5}{8} + 6\frac{5}{6}$

15. $16\frac{2}{5} + 13\frac{3}{4}$

16. $13\frac{4}{15} + 12\frac{3}{5}$

17. $16\frac{2}{5} + 8\frac{1}{10}$

18. $12\frac{3}{8} + 10\frac{3}{4}$

19. $4\frac{4}{9} + 5\frac{5}{12}$

20. $18\frac{2}{3} + 12\frac{8}{9}$

21. $10\frac{6}{7} + 5\frac{2}{5}$

22. $15\frac{3}{4} + 8\frac{5}{8}$

23. $24\frac{1}{2} + 12\frac{3}{4}$

24. $20\frac{2}{9} + 8\frac{1}{12}$

25. $8\frac{2}{11} + 6\frac{1}{2} =$

26. $9\frac{5}{9} + 10\frac{5}{12} =$

27. $6\frac{4}{9} + 8\frac{7}{15} =$

28. $12\frac{4}{15} + 5\frac{7}{12} =$

29. $14\frac{4}{9} + 10\frac{2}{3} =$

30. $19\frac{2}{7} + 12\frac{5}{21} =$

Name ______________________ Date ____________ Period ________

Subtracting Fractions

To subtract fractions with like denominators, subtract the numerators. Write the difference over the common denominator. Simplify the difference if possible.

Example 1 **Subtract:** $\frac{3}{4} - \frac{1}{4}$.

Step 1	Step 2
Subtract the numerators. Write the difference over the like denominator. $\frac{3}{4} - \frac{1}{4} = \frac{3-1}{4}$ or $\frac{2}{4}$	Simplify the difference. $\frac{2}{4} \overset{\div 2}{=} \frac{1}{2}$ *The GCF of 2 and 4 is 2.*

To subtract fractions with unlike denominators, rename the fractions with a common denominator. Then subtract the fractions.

Example 2 **Subtract:** $\frac{7}{10} - \frac{2}{5}$.

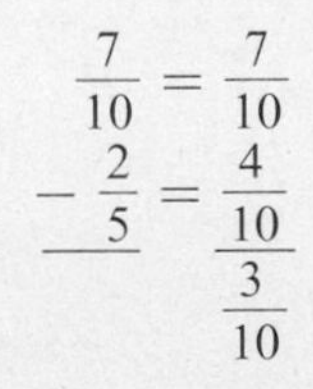

Use 10 for the common denominator.

Subtract. Write each difference in simplest form.

1. $\frac{5}{6} - \frac{4}{6}$

2. $\frac{9}{10} - \frac{3}{10}$

3. $\frac{9}{16} - \frac{3}{16}$

4. $\frac{11}{12} - \frac{3}{12}$

SKILL 30

Name ______________________ Date ____________ Period ______

Subtracting Fractions *(continued)*

Subtract. Write each difference in simplest form.

5. $\frac{11}{14} - \frac{5}{14}$

6. $\frac{8}{9} - \frac{2}{9}$

7. $\frac{5}{6} - \frac{1}{3}$

8. $\frac{11}{12} - \frac{3}{4}$

9. $\frac{9}{10} - \frac{2}{5}$

10. $\frac{5}{7} - \frac{3}{14}$

11. $\frac{20}{21} - \frac{5}{14}$

12. $\frac{9}{14} - \frac{1}{2}$

13. $\frac{11}{15} - \frac{3}{10}$

14. $\frac{5}{6} - \frac{1}{12}$

15. $\frac{7}{18} - \frac{1}{6}$

16. $\frac{9}{20} - \frac{1}{8}$

17. $\frac{7}{12} - \frac{2}{9}$

18. $\frac{13}{18} - \frac{5}{12}$

19. $\frac{9}{16} - \frac{1}{6}$

20. $\frac{17}{24} - \frac{3}{10}$

Name ____________ Date ________ Period ______

Subtracting Fractions and Mixed Numbers

To subtract fractions and mixed numbers, first rename each fraction by finding the LCD if necessary. Then rename, if necessary, to subtract. Next subtract the fractions and then the whole numbers. Rename and simplify if necessary.

Example 1 **Find $4\frac{2}{5} - 1\frac{9}{10}$.**

Step 1	Step 2	Step 3
Rename each fraction finding the LCD if necessary. $4\frac{2}{5} \rightarrow 4\frac{4}{10}$; $-1\frac{9}{10} \rightarrow -1\frac{9}{10}$	Rename if necessary to subtract. $4\frac{4}{10} = 3 + \frac{10}{10} + \frac{4}{10} = 3\frac{14}{10}$; $4\frac{4}{10} \rightarrow 3\frac{14}{10}$; $-1\frac{9}{10} \rightarrow -1\frac{9}{10}$	Subtract and simplify if necessary. $3\frac{14}{10} - 1\frac{9}{10} = 2\frac{5}{10}$ or $2\frac{1}{2}$

Example 2 **Find $6 - 3\frac{1}{6}$.**

$$6 \rightarrow 5\frac{6}{6}$$
$$-3\frac{1}{6} \rightarrow -3\frac{1}{6}$$
$$2\frac{5}{6}$$

Subtract. Write each difference in simplest form.

1. $14\frac{2}{3} - 12$
2. $10 - 4\frac{3}{4}$
3. $7\frac{7}{9} - 3\frac{4}{9}$
4. $8\frac{1}{3} - 4\frac{2}{3}$
5. $15\frac{1}{4} - 5\frac{1}{2}$
6. $16\frac{3}{8} - 2\frac{5}{6}$
7. $14\frac{3}{7} - 10\frac{1}{2}$
8. $18\frac{3}{10} - 7\frac{4}{5}$

SKILL 31

Name ______________________ Date ____________ Period ______

Subtracting Fractions and Mixed Numbers *(continued)*

Subtract. Write each difference in simplest form.

9. $8\frac{1}{5} - 2\frac{3}{5}$

10. $6 - 3\frac{2}{7}$

11. $9\frac{5}{12} - 3\frac{3}{4}$

12. $16\frac{2}{9} - 2\frac{2}{3}$

13. $23\frac{1}{2} - 15\frac{1}{4}$

14. $13\frac{2}{15} - 8\frac{1}{5}$

15. $16\frac{3}{8} - 14\frac{3}{4}$

16. $19\frac{1}{6} - 4\frac{2}{3}$

17. $9\frac{2}{9} - 1\frac{1}{18}$

18. $7 - 2\frac{4}{7}$

19. $16\frac{3}{4} - 5\frac{11}{12}$

20. $12\frac{1}{3} - 10\frac{3}{4}$

21. $26\frac{1}{4} - 15\frac{3}{5}$

22. $14\frac{1}{9} - 8\frac{2}{3}$

23. $15\frac{1}{8} - 6\frac{1}{4}$

24. $18\frac{1}{2} - 9\frac{7}{8}$

25. $6\frac{3}{11} - 5\frac{1}{3}$

26. $12\frac{5}{7} - 6\frac{1}{2}$

27. $8\frac{2}{9} - 1\frac{7}{12}$

28. $14\frac{3}{10} - 6\frac{4}{5}$

29. $12\frac{5}{6} - 10\frac{2}{3}$

30. $21\frac{2}{5} - 18\frac{7}{15}$

Name ______________________ Date __________ Period ______

Changing Fractions to Decimals

A fraction is another way of writing a division problem. To change a fraction to a decimal, divide the numerator by the denominator.

$\frac{1}{10}$

$\frac{1}{100}$

Examples ***1*** **About $\frac{1}{20}$ of the heat in a house is lost through the doors. Write this fraction as a decimal.**

$\frac{1}{20}$ means $1 \div 20$ or $20\overline{)1}$.

$$20\overline{)1.00}\quad 0.05$$

So, $\frac{1}{20} = 0.05$.

2 **Express $\frac{1}{3}$ as a decimal.**

$$3\overline{)1.00}\quad 0.33...$$

$\frac{1}{3} = 0.33...$ or $0.\overline{3}$ *The bar status shows that 3 repeats.*

Express each fraction as a decimal. Use bar notation if necessary.

1. $\frac{4}{25}$ **2.** $\frac{3}{5}$ **3.** $\frac{7}{20}$ **4.** $\frac{3}{50}$

5. $\frac{9}{10}$ **6.** $\frac{7}{8}$ **7.** $\frac{1}{3}$ **8.** $\frac{14}{16}$

9. $\frac{20}{30}$ **10.** $\frac{5}{9}$ **11.** $\frac{19}{20}$ **12.** $\frac{5}{200}$

SKILL 32

Name ______________ Date ________ Period ______

Changing Fractions to Decimals *(continued)*

Express each fraction as a decimal. Use bar notation if necessary.

13. $\frac{10}{50}$ **14.** $\frac{13}{20}$ **15.** $\frac{5}{6}$ **16.** $\frac{4}{5}$

17. $\frac{7}{10}$ **18.** $\frac{13}{40}$ **19.** $\frac{39}{50}$ **20.** $\frac{2}{25}$

21. $\frac{7}{16}$ **22.** $\frac{34}{125}$ **23.** $\frac{16}{25}$ **24.** $\frac{99}{100}$

25. $\frac{17}{20}$ **26.** $\frac{3}{150}$ **27.** $\frac{3}{8}$ **28.** $\frac{2}{3}$

A mill is a unit of money that is used in assessing taxes. One mill is equal to $\frac{1}{10}$ of a cent or $\frac{1}{1000}$ of a dollar.

29. Money is usually written using decimals. Express each fraction above as a decimal using the correct money symbol.

30. Find the number of cents and the number of dollars equal to 375 mills.

31. Find the number of cents and the number of dollars equal to 775 mills.

32. Find the number of cents and the number of dollars equal to 1,000 mills.

Name ______________________ Date __________ Period ______

Writing Decimals as Fractions

To write a terminating decimal as a fraction, write the digits to the right of the decimal point over a power of ten. The power of ten is determined by the place-value position of the last digit in the decimal. For example, if the last digit is in the hundredths place, use 100. If the last digit is in the thousandths place, use 1000.

Example ***1*** **Write 0.375 as a fraction.**

Since the last digit, 5, is in the thousandths place, write 375 over 1000. Then simplify.

$0.375 = \frac{375}{1000}$ or $\frac{3}{8}$

Repeating decimals can also be written as fractions using the method shown below.

Example ***2*** **Write 0.555... as a fraction.**

Let $N = 0.555\ldots$. Then $10N = 5.555\ldots$.
Subtract N from $10N$ to eliminate the repeating part.

$$\begin{aligned} 10N &= 5.555\ldots \\ -\ N &= 0.555\ldots \\ \hline 9N &= 5 \\ N &= \frac{5}{9} \end{aligned}$$

So, $0.555\ldots = \frac{5}{9}$.

Write each decimal as a fraction.

1. 0.525 **2.** 0.45 **3.** 0.333…

4. 0.43 **5.** 0.8 **6.** 0.1212…

7. 0.345 **8.** 0.1862 **9.** 0.4555…

SKILL 33

Name ______________________ Date ____________ Period ________

Writing Decimals as Fractions *(continued)*

Write each decimal as a fraction.

10. 0.456

11. 0.32

12. 0.222…

13. 0.35

14. 0.48

15. 0.955

16. 0.8222…

17. 0.4545…

18. 0.444…

19. 0.565

20. 0.435

21. 0.552

22. 0.855

23. 0.842

24. 0.944

25. 0.732

26. 0.245

27. 0.485

28. 0.666…

29. 0.8585…

30. 0.9655

SKILL 34

Name ______________________ Date __________ Period ______

Writing Decimals as Percents

To express a decimal as a percent, first express the decimal as a fraction with a denominator of 100. Then express the fraction as a percent.

Examples **Express each decimal as a percent.**

1 $\mathbf{0.09} = \frac{9}{100}$

$= 9\%$

2 $\mathbf{0.005} = \frac{5}{1000}$

$= \frac{0.5}{100}$

$= 0.5\%$

3 $\mathbf{1.8} = \frac{18}{10}$

$= \frac{180}{100}$

$= 180\%$

A shortcut to writing a decimal as a percent is to move the decimal point two places to the right and add a percent sign (%).

Examples **Express each decimal as a percent.**

4 **0.25**

0.25 = 0.25%

= 25%

5 **0.9**

0.9 = 0.90%

= 90%

Express each decimal as a percent.

1. 0.66 **2.** 0.08 **3.** 0.75 **4.** 0.001

5. 1.19 **6.** 0.72 **7.** 0.136 **8.** 4.02

9. 0.18 **8.** 0.36 **11.** 0.09 **12.** 0.2

13. 0.625 **14.** 0.007 **15.** 1.4 **16.** 0.093

SKILL 34

Name ______________ Date __________ Period ______

Writing Decimals as Percents *(continued)*

Express each decimal as a percent.

17. 0.8 **18.** 0.54 **19.** 3.75 **20.** 0.02

21. 0.258 **22.** 0.016 **23.** 0.49 **24.** 0.003

25. 0.96 **26.** 0.52 **27.** 0.15 **28.** 0.008

29. 3.62 **30.** 0.623 **31.** 0.035 **32.** 7.08

33. 0.5 **34.** 0.97 **35.** 0.6 **36.** 0.425

37. 0.08 **38.** 2.5 **39.** 0.001 **40.** 0.074

41. 0.345 **42.** 0.19 **43.** 0.062 **44.** 0.19

45. 0.005 **46.** 0.37 **47.** 0.8 **48.** 0.04

Name ______________________ Date __________ Period ______

Writing Percents as Decimals

To express a percent as a decimal, divide by 100 and write as a decimal.

Examples **Express each percent as a decimal.**

1 **56%**

$56\% = \frac{56}{100}$

$= 0.56$

2 **3.4%**

$3.4\% = \frac{3.4}{100}$

$= \frac{34}{1000}$

$= 0.034$

A shortcut to writing a percent as a decimal is to move the decimal point two places to the left and drop the percent sign.

Examples **Express each percent as a decimal.**

3 **18%**

18% = 18.

= 0.18

4 **0.5%**

0.5% = 000.5

= 0.005

Express each percent as a decimal.

1. 45% **2.** 91% **3.** 24.5% **4.** 8.37%

5. 13% **6.** 6% **7.** 76.5% **8.** 1.22%

9. 14.5% **10.** 26% **11.** 1.8% **12.** 80%

SKILL 35

Name ______________________ Date __________ Period ______

Writing Percents as Decimals *(continued)*

Express each percent as a decimal.

13. 8%

14. 32%

15. 15%

16. 15.7%

17. 16.23%

18. 2.01%

19. 3.2%

20. 80%

21. 1.32%

22. 21%

23. 25%

24. 13%

25. 4%

26. 40%

27. 62.5%

28. 30%

29. 60.3%

30. 12.3%

31. 10.25%

21. 8.6%

33. 12.15%

34. 102%

35. 450.5%

36. 175%

37. 0.05%

38. 0.25%

39. 0.105%

40. 14.36%

41. 2.18%

42. 38.65%

Name ______________________ Date ____________ Period ________

Writing Fractions as Percents

To express a fraction as a percent, first set up a proportion. Then solve the proportion using cross products.

Example **Express $\frac{13}{20}$ as a percent.**

$\frac{13}{20} = \frac{k}{100}$ *Set up a proportion.*

$13 \times 100 = 20 \times k$ *Find the cross products.*

$1300 = 20k$

$1300 \div 20 = 20k \div 20$ *Divide each side by 20.*

$65 = k$

$\frac{13}{20} = \frac{65}{100}$ or 65%

Express each shaded section as a fraction and as a percent.

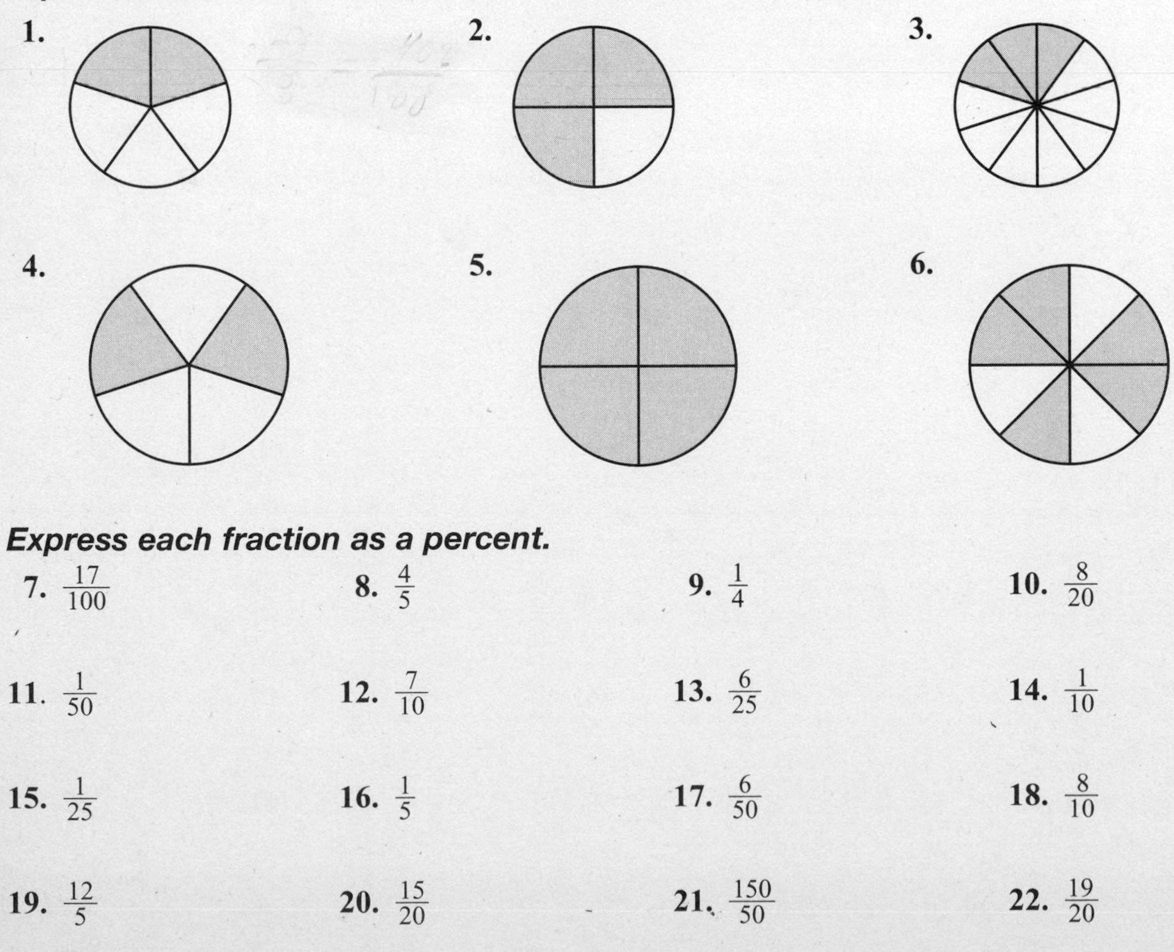

Express each fraction as a percent.

7. $\frac{17}{100}$ 8. $\frac{4}{5}$ 9. $\frac{1}{4}$ 10. $\frac{8}{20}$

11. $\frac{1}{50}$ 12. $\frac{7}{10}$ 13. $\frac{6}{25}$ 14. $\frac{1}{10}$

15. $\frac{1}{25}$ 16. $\frac{1}{5}$ 17. $\frac{6}{50}$ 18. $\frac{8}{10}$

19. $\frac{12}{5}$ 20. $\frac{15}{20}$ 21. $\frac{150}{50}$ 22. $\frac{19}{20}$

Name ______________________ Date ____________ Period ________

Writing Fractions as Percents *(continued)*

Use a 10 × 10 grid to shade the amount stated in each fraction. Then express each fraction as a percent.

23. $\frac{1}{10}$

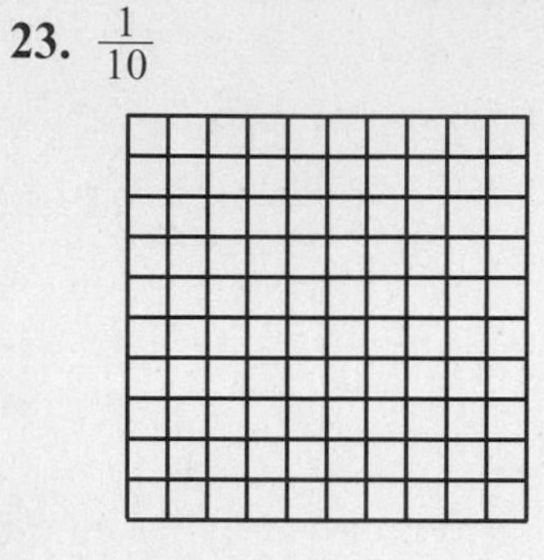

24. $\frac{1}{20}$

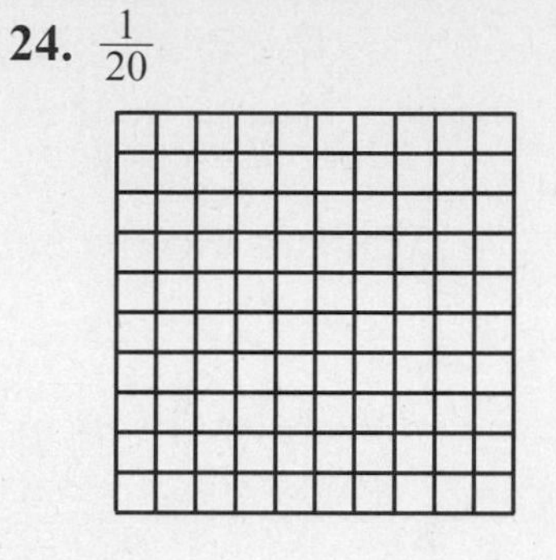

25. $\frac{1}{50}$

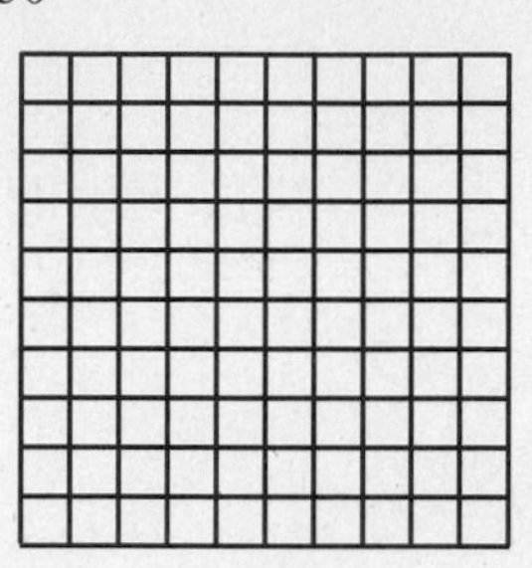

Express each fraction as a percent.

26. $\frac{47}{100}$	**27.** $\frac{8}{25}$	**28.** $\frac{9}{12}$	**29.** $\frac{13}{50}$
30. $\frac{11}{20}$	**31.** $\frac{7}{5}$	**32.** $\frac{3}{100}$	**33.** $\frac{21}{25}$
34. $\frac{3}{10}$	**35.** $\frac{3}{20}$	**36.** $\frac{31}{50}$	**37.** $\frac{5}{4}$
38. $\frac{6}{10}$	**39.** $\frac{15}{5}$	**40.** $\frac{12}{50}$	**41.** $\frac{1}{20}$
42. $\frac{17}{20}$	**43.** $\frac{152}{50}$	**44.** $\frac{400}{100}$	**45.** $\frac{30}{25}$
46. $\frac{9}{10}$	**47.** $\frac{49}{50}$	**48.** $\frac{24}{25}$	**49.** $\frac{40}{20}$
50. $\frac{5}{15}$	**51.** $\frac{12}{20}$	**52.** $\frac{18}{10}$	**53.** $\frac{1000}{100}$
54. $\frac{13}{20}$	**55.** $\frac{215}{50}$	**56.** $\frac{25}{20}$	**57.** $\frac{8}{5}$
58. $\frac{16}{10}$	**59.** $\frac{43}{50}$	**60.** $\frac{75}{25}$	**61.** $\frac{22}{20}$

SKILL 37

Name ______________________ Date __________ Period ______

Writing Percents as Fractions

To express a percent as a fraction, divide by 100 and simplify.

Examples **Express each percent as a fraction.**

1 **65%**

$65\% = \frac{65}{100}$

$= \frac{13}{20}$

2 **2.5%**

$2.5\% = \frac{2.5}{100}$

$= \frac{25}{1000}$

$= \frac{1}{40}$

Express each percent as a fraction.

1. 45%

2. 91%

3. 24.5%

4. 8%

5. 32%

6. 15%

7. 15.7%

8. 16.23%

9. 2.01%

10. 3.2%

11. 80%

12. 1.32%

13. 21%

14. 25%

15. 13%

SKILL 37

Name ____________________ Date __________ Period ______

Writing Percents as Fractions *(continued)*

Express each percent as a fraction.

16. 4%

17. 40%

18. 62.5%

19. 30%

20. 60.3%

21. 12.3%

22. 15%

23. 32%

24. 67%

25. 62.8%

26. 18%

27. 23%

28. 70%

29. 1.5%

30. 3.2%

31. 1.82%

32. 14.8%

33. 16%

34. 120%

35. 18.5%

36. 255%

37. 100.5%

38. 1.255%

39. 6.8%

40. 0.09%

41. 45.45%

42. 50.15%

Name ______________________ Date __________ Period ______

Comparing and Ordering Rational Numbers

To compare fractions, write each fraction as a decimal. Then compare the decimals.

***Example* 1** **Compare $\frac{2}{3}$ and $\frac{3}{5}$.**

$\frac{2}{3} = 0.6666666667$

$\frac{3}{5} = 0.6$

Since $0.6666666667 > 0.6$, $\frac{2}{3} > \frac{3}{5}$.

To compare percents, compare the numbers without the percent sign.

***Example* 2** **Compare 15% and 17.5%.**

Since $15 < 17.5$, $15\% < 17.5\%$.

Fill in each ◯ with $<$, $>$, or $=$ to make a true sentence.

1. $\frac{2}{7}$ ◯ $\frac{3}{8}$
2. $\frac{3}{11}$ ◯ $\frac{1}{5}$
3. $\frac{11}{21}$ ◯ $\frac{9}{16}$
4. $\frac{14}{21}$ ◯ $\frac{10}{15}$
5. $\frac{25}{27}$ ◯ $\frac{17}{19}$
6. $\frac{3}{10}$ ◯ $\frac{4}{9}$
7. $1\frac{7}{8}$ ◯ $2\frac{4}{5}$
8. $3\frac{7}{9}$ ◯ $3\frac{6}{7}$
9. $5\frac{10}{19}$ ◯ $5\frac{15}{24}$
10. 14% ◯ 12.5%
11. 5% ◯ 8%
12. 0.04% ◯ 0.25%
13. 250% ◯ 126%
14. 16.6% ◯ 10%
15. 75.8% ◯ 75.9%

SKILL 38

Name ____________________ Date __________ Period ______

Comparing and Ordering Rational Numbers *(continued)*

Write each set of fractions in order from least to greatest.

16. $\frac{3}{5}, \frac{7}{9}, \frac{4}{5}, \frac{1}{2}$

17. $\frac{3}{8}, \frac{2}{7}, \frac{8}{11}, \frac{5}{16}$

18. $\frac{9}{14}, \frac{6}{7}, \frac{3}{4}, \frac{12}{19}$

19. $\frac{11}{23}, \frac{19}{27}, \frac{7}{10}, \frac{15}{17}$

The Pittsburgh Pirates have won 14 out of 21 games, and the New York Mets have won 15 out of 23 games. Use this information to answer Exercises 20–23.

20. Which team has the better record?

21. Suppose the Pirates win 2 of their next three games and the Mets win all of their next 3 games. Which team has the better record?

22. Suppose the Pirates went on to win 21 games after playing 30 games. Is their record better now than it was before? Explain.

23. Suppose the Mets went on to win 16 games after playing 30 games. Is their record better now than it was before? Explain.

24. Larry has $\frac{5}{6}$ yard of material. Does he have enough to make a vest that requires $\frac{3}{4}$ yard of material? Explain.

Name ______________________ Date ____________ Period ________

Length in the Customary System

Length
1 foot (ft) = 12 inches (in.)
1 yard (yd) = 3 feet or 36 inches
1 mile (mi) = 5280 feet or 1760 yards

Example 1 **Draw a line segment measuring $3\frac{3}{8}$ inches.**

Use a ruler divided in eighths.

Find $3\frac{3}{8}$ on the ruler.

Draw the line segment from 0 to $3\frac{3}{8}$.

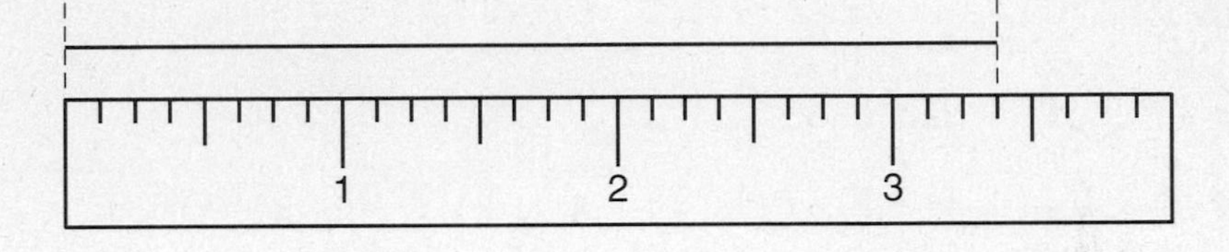

To change from a smaller unit to a larger unit, divide.
To change from a larger unit to a smaller unit, multiply.

Examples **2** **3 ft = ________ in.** *1 ft = 12 in., so multiply by 12.*

$3 \times 12 = 36$

3 ft = 36 in.

3 **9 ft = ________ yd** *1 yd = 3 ft, so divide by 3.*

$9 \div 3 = 3$

9 ft = 3 yd

Draw a line segment of each length.

1. $1\frac{1}{2}$ inches

2. $1\frac{1}{8}$ inches

3. $1\frac{1}{4}$ inches

4. $\frac{3}{4}$ inch

5. $1\frac{3}{8}$ inches

6. $\frac{5}{8}$ inches

7. $3\frac{1}{2}$ inches

8. $\frac{3}{8}$ inches

9. $1\frac{3}{4}$ inch

10. $2\frac{1}{4}$ inches

11. $2\frac{5}{8}$ inches

12. $3\frac{1}{8}$ inches

SKILL 39

Name ______________________ Date ____________ Period ________

Length in the Customary System *(continued)*

Complete.

13. 5 ft = ________ in.

14. 2 mi = ________ ft

15. 12 yd = ________ ft

16. 24 in. = ________ yd

17. 48 in. = ________ ft

18. 3520 yd = ________ mi

19. 72 in. = ________ yd

20. 30 in. = ________ ft

21. 4 mi = ________ ft

22. 90 in. = ________ yd

23. 60 in. = ________ yd

24. 6 mi = ________ yd

25. 6.5 ft = ________ in.

26. 15 ft = ________ yd

27. 9 yd = ________ in.

28. 12 ft = ________ in.

29. 7920 ft = ________ mi

30. 16 ft = ________ in.

Name ______________________ Date __________ Period ______

Capacity in the Customary System

Capacity
1 cup (c) = 8 fluid ounces (fl oz)
1 pint (pt) = 2 cups
1 quart (qt) = 2 pints
1 gallon (gal) = 4 quarts

To change from one customary unit of capacity to another, you either multiply or divide.
When changing from a smaller unit to a larger unit, divide.
When changing from a larger unit to a smaller unit, multiply.

Examples ***1*** **12 qt = ________ pt**

You are changing from a larger unit (qt) to a smaller unit (pt), so multiply.

$12 \times 2 = 24$ — *Since 1qt = 2 pt, multiply by 2.*

12 qt = 24 pt

2 **8 pt = ________ gal**

You are changing from a smaller unit (pt) to a larger unit (gal), so divide.

$8 \div 2 = 4$ — *Divide by 2 to change pints to quarts.*

$4 \div 4 = 1$ — *Divide by 4 to change quarts to gallons.*

8 pt = 1 gal

Complete.

1. 8 c = ______ fl oz

2. 8 qt = ______ gal

3. 16 pt = ______ qt

4. 5 c = ______ pt

5. 16 qt = ______ pt

6. 18 c = ______ qt

7. 8 gal = ______ qt

8. 16 gal = ______ qt

SKILL 40

Name ____________________ Date __________ Period ______

Capacity in the Customary System *(continued)*

Complete.

9. 16 fl oz = ______ c

10. 16 pt = ______ c

11. 3 qt = ______ pt

12. 5 gal = ______ qt

13. 15 pt = ______ qt

14. 12pt = ______ c

15. 16 c = ______ fl oz

16. 10 pt = ______ qt

17. 3 qt = ______ c

18. 12 c = ______ fl oz

19. 64 pt = ______ gal

20. 4 gal = ______ c

21. 1 qt = ______ fl oz

22. 5 c = ______ fl oz

23. 17 c = ______ pt

24. 6 qt = ______ gal

25. 2.5 gal = ______ qt

26. $3\frac{1}{2}$ gal = ______ qt

27. 16 qt = ______ gal

28. 80 fl oz = ______ pt

29. 16 qt = ______ c

30. 8 c = ______ qt

31. A recipe calls for 3 cups of milk How many fluid ounces of milk are need for the recipe?

32. Jenna bought 64 fl oz of juice. How many quarts of juice did she buy?

SKILL 41

Name ________________ Date ________ Period ______

Weight in the Customary System

Weight
1 pound (lb) = 16 ounces (oz)
1 ton (T) = 2000 pounds

To change from one customary unit of weight to another, you either multiply or divide.
When changing from a smaller unit to a larger unit, divide.
When changing from a larger unit to a smaller unit, multiply.

Examples ***1*** $\mathbf{10\frac{1}{2}}$ **lb = ________ oz**

You are changing from a larger unit (lb) to a smaller unit (oz), so multiply.

$10\frac{1}{2} \times 16 = \frac{21}{\cancel{2}_1} \times \frac{\cancel{16}^8}{1} = \frac{168}{1}$ or 168

Since 1 pound =16 ounces, multiply by 16.

$10\frac{1}{2}$ lb = 168 oz

2 **32 oz = ________ lb**

You are changing from a smaller unit (oz) to a larger unit (lb), so divide.

32 ÷ 16 = 2

Divide by 16 to change ounces to pounds.

32 oz = 2 lb

Complete.

1. 2 T = ____________ lb

2. 8500 lb = ____________ T

3. 24 oz = ____________ lb

4. 4 lb = ____________ oz

5. $3\frac{1}{2}$ lb = ____________ oz

6. 2500 lb = ____________ T

7. 10 lb = ____________ oz

8. 1 T = ____________ oz

SKILL 41

Name ______________________ Date ____________ Period ________

Weight in the Customary System *(continued)*

Complete.

9. 256 oz = ____________ lb

10. 16 lb = ____________ oz

11. 3 T = ____________ lb

12. 7 T = ____________ lb

13. 12,000 lb = ____________ T

14. 12 oz = ____________ lb

15. 16 T = ____________ lb

16. 10 T = ____________ oz

17. 3 lb = ____________ oz

18. 12 oz = ____________ lb

19. 64 oz = ____________ lb

20. 4 oz = ____________ lb

21. 2.5 T = ____________ lb

22. 5 lb = ____________ oz

23. 17 oz = ____________ lb

24. 6 oz = ____________ lb

25. $\frac{1}{5}$ T = ____________ lb

26. $3\frac{1}{2}$ T = ____________ oz

27. 6.5 T = ____________ lb

28. 500 lb = ____________ T

29. 20 lb = ____________ oz

30. 2.25 T = ____________ lb

31. A recipe calls for 3 ounces of butter How many pounds of butter are needed for the recipe?

32. Jenna bought 64 ounces of bananas. How many pounds of bananas did she buy?

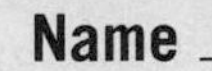

Name ______________________ Date __________ Period ______

Length in the Metric System

Length
1 centimeter (cm) = 10 millimeters (mm)
1 meter (m) = 100 centimeters
1 meter = 1000 millimeters
1 kilometer (km) = 1000 meters

To change from one metric unit of length to another, you either multiply or divide.
When changing from a smaller unit to a larger unit, divide.
When changing from a larger unit to a smaller unit, multiply.

Examples ***1*** **3 m = ________ mm**

You are changing from a larger unit (m) to a smaller unit (mm), so multiply.

$3 \times 1000 = 3000$

Since 1 m = 1000 mm, multiply by 1000. Move the decimal point 3 places to the right.

3 m = 3000 mm

2 **5000 m = ________ km**

You are changing from a smaller unit (m) to a larger unit (km), so divide.

$5000 \div 1000 = 5.000$

Since 1000 meters = 1 kilometer, divide by 1000. Move the decimal point 3 places to the left.

5000 m = 5 km

Complete.

1. 300 mm = ____________ cm

2. 2000 m = ____________ km

3. 60 cm = ____________ m

4. 1500 m = ____________ km

5. 6 km = ____________ m

6. 8 km = ____________ cm

7. 80 mm = ____________ cm

8. 160 cm = ____________ m

SKILL 42

Name ____________________ Date __________ Period ______

Length In the Metric System *(continued)*

Complete.

9. 2000 mm = ____________ cm

10. 2 m = ____________ cm

11. 300 mm = ____________ cm

12. 7 cm = ____________ mm

13. 160 cm = ____________ mm

14. 20 km = ____________ m

15. 3000 cm = ____________ m

16. 24,000 mm = ____________ m

17. 2000 km = ____________ m

18. 42 cm = ____________ mm

19. 4100 cm = ____________ m

20. 8700 cm = ____________ m

21. 42,000 m = ____________ km

22. 4 km = ____________ m

23. 8 m = ____________ cm

24. 50 cm = ____________ mm

25. 16.3 mm = ____________ cm

26. 4.1 km = ____________ m

27. 15.5 cm = ____________ mm

28. 160 km = ____________ m

29. A napkin is 37 centimeters long. How many millimeters is this?

30. A race is 80,000 meters long. How long is the race in kilometers?

Name ______________________ Date ____________ Period ________

Capacity in the Metric System

Capacity
1 liter (L) = 1000 milliliters (mL)
1 kiloliter (kL) = 1000 liters

To change from one metric unit of capacity to another, you either multiply or divide.
When changing from a smaller unit to a larger unit, divide.
When changing from a larger unit to a smaller unit, multiply.

Examples ***1*** **1325 mL = ________ L**

You are changing from a smaller unit (mL) to a larger unit (L), so divide.

1325 ÷ 1000 = 1.325

Since 1 mL = 1000 L, divide by 1000. Move the decimal point 3 places to the left.

1325 mL = 1.325 L

2 **2 kL = ________ L**

You are changing from a larger unit (kL) to a smaller unit (L), so multiply.

2 × 1000 = 2000

Since 1 kL = 1000 L, multiply by 1000. Move the decimal point 3 places to the right.

2 kL = 2000 L

Complete.

1. 76 mL = ____________ L

2. 1800 L = ____________ kL

3. 140 L = ____________ mL

4. 7500 L = ____________ mL

5. 8.2 kL = ____________ L

6. 140 L = ____________ kL

7. 6000 mL = ____________ L

8. 400 kL = ____________ L

SKILL 43

Name ______________________ Date __________ Period ______

Capacity in the Metric System *(continued)*

Complete.

9. 5 kL = ____________ L

10. 2000 L = ____________ kL

11. 4 L = ____________ mL

12. 1400 L = ____________ kL

13. 3250 mL = ____________ L

14. 3.4 kL = ____________ L

15. 750 L = ____________ kL

16. 940 mL = ____________ L

17. 12 L = ____________ mL

18. 3400 mL = ____________ L

19. 86 kL = ____________ L

20. 8 L = ____________ mL

21. 36 kL = ____________ L

22. 850 L = ____________ kL

23. 2.4 L = ____________ mL

24. 3.8 kL = ____________ L

25. 5.35 L = ____________ mL

26. 10.6 kL = ____________ L

27. 180 L = ____________ kL

28. 1400 mL = ____________ L

29. Karen uses 2 L of liquid in her punch recipe. How many mL does she use?

30. José brought home a soft drink bottle that contained 2000 milliliters of liquid. What is the capacity in liters?

SKILL 44

Name ______________ Date ________ Period ______

Mass in the Metric System

Mass
1 gram (g) = 1000 milligrams (mg)
1 kilogerams (kg) = 1000 grams

To change from one metric unit of mass to another, you either multiply or divide.
When changing from a smaller unit to a larger unit, divide.
When changing from a larger unit to a smaller unit, multiply.

Examples ***1*** **1325 mg = ________ g**

You are changing from a smaller unit (mg) to a larger unit (g), so divide.

1325 ÷ 1000 = 1.325

Since 1 mg = 1000 g, divide by 1000. Move the decimal point 3 places to the left.

1325 mg = 1.325 g

2 **76 kg = ________ g**

You are changing from a larger unit (kg) to a smaller unit (g), so multiply.

76 × 1000 = 76,000

Since 1 kg = 1000 g, multiply by 1000. Move the decimal point 3 places to the right.

76 kg = 76,000 g

Complete.

1. 180 mg = ____________ g

2. 1600 g = ____________ kg

3. 1500 kg = ____________ g

4. 700 mg = ____________ g

5. 8000 g = ____________ mg

6. 450 kg = ____________ g

7. 820 g = ____________ kg

8. 4630 mg = ____________ g

SKILL 44

Name ______________ Date ________ Period ______

Mass in the Metric System *(continued)*

Complete.

9. 5 kg = __________ g

10. 2000 g = __________ kg

11. 4 g = __________ mg

12. 1400 g = __________ kg

13. 3250 mg = __________ g

14. 3.4 kg = __________ g

15. 750 g = __________ kg

16. 940 mg = __________ g

17. 12 g = __________ mg

18. 3400 mg = __________ g

19. 86 kg = __________ g

20. 8 g = __________ mg

21. 36 kg = __________ g

22. 850 g = __________ kg

23. 2.4 g = __________ mg

24. 3.8 kg = __________ g

25. 5.35 g = __________ mg

26. 10.6 kg = __________ g

27. 86 mg = __________ g

28. 140 kg = __________ g

29. Mr. Chang's truck can carry a payload of 11 kilograms. What is the payload in grams?

30. Jana weighed her dog at 20 kg. What is the weight of her dog in mg?

Name ______________________ Date ____________ Period ________

Converting Customary Units to Metric Units

You can use the following chart to convert customary units to metric units.

Customary Unit / Metric Unit
1 in. = 2.54 cm
1 ft = 30.48 cm or 0.3048 m
1 yd ≈ 0.914 m
1 mi ≈ 1.609 km
1 oz = 28.350 g
1 lb ≈ 454 g or 0.454 kg
1 fl oz = 29.574 mL
1 qt = 0.946 L
1 gal = 3.785 L

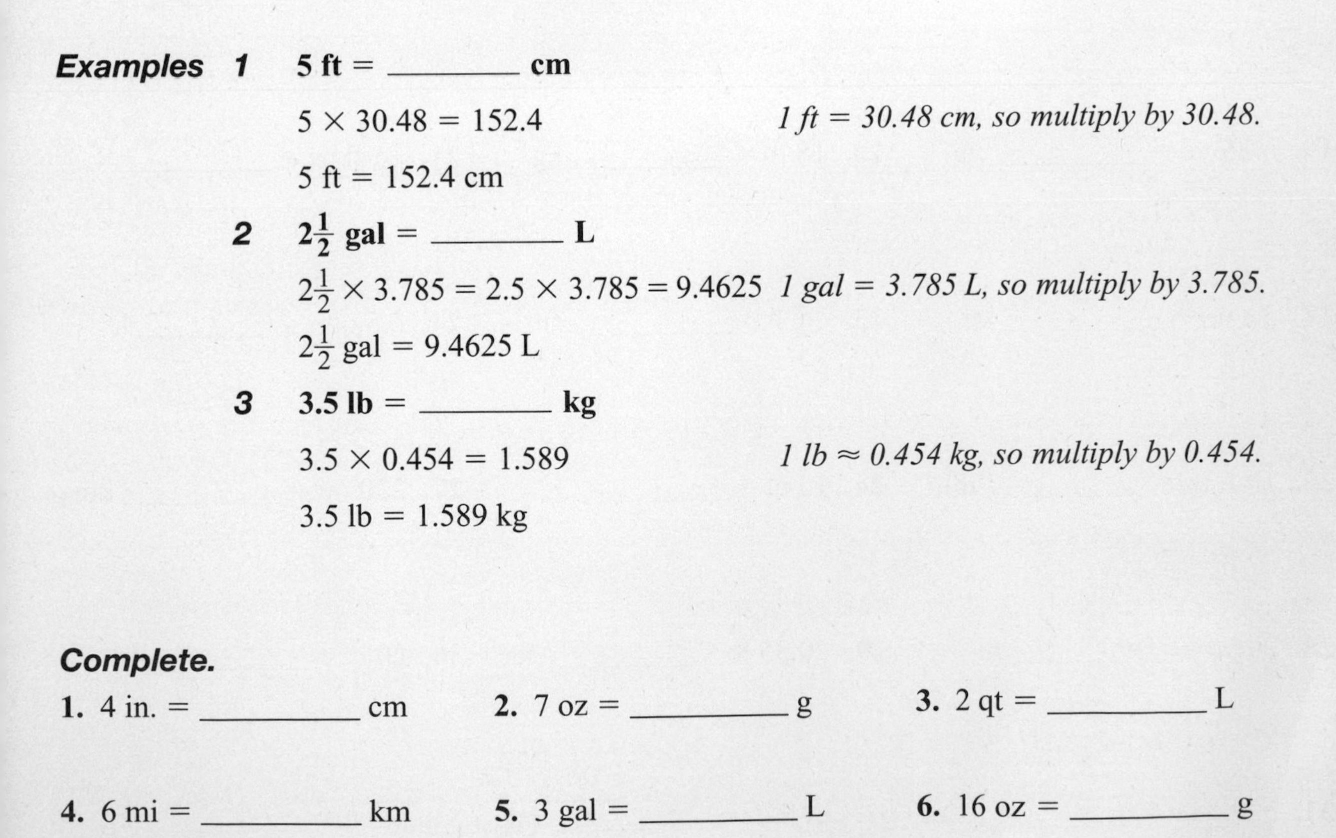

Examples ***1*** **5 ft = ________ cm**

$5 \times 30.48 = 152.4$ *1 ft = 30.48 cm, so multiply by 30.48.*

5 ft = 152.4 cm

2 **$2\frac{1}{2}$ gal = ________ L**

$2\frac{1}{2} \times 3.785 = 2.5 \times 3.785 = 9.4625$ *1 gal = 3.785 L, so multiply by 3.785.*

$2\frac{1}{2}$ gal = 9.4625 L

3 **3.5 lb = ________ kg**

$3.5 \times 0.454 = 1.589$ *1 lb ≈ 0.454 kg, so multiply by 0.454.*

3.5 lb = 1.589 kg

Complete.

1. 4 in. = ____________ cm **2.** 7 oz = ____________ g **3.** 2 qt = ____________ L

4. 6 mi = ____________ km **5.** 3 gal = ____________ L **6.** 16 oz = ____________ g

SKILL 45

Name ______________ Date __________ Period ______

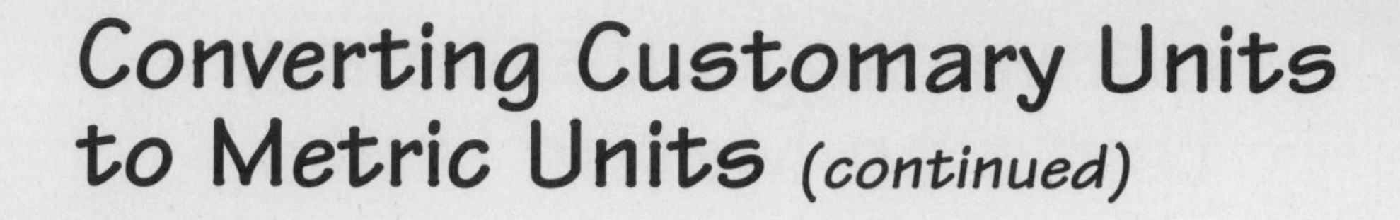

Converting Customary Units to Metric Units *(continued)*

Complete.

7. 12 fl oz = __________ mL **8.** 5 lb = __________ g **9.** 3 yd = __________ m

10. 1.5 in. = __________ cm **11.** 4 ft = __________ m **12.** 5 qt = __________ L

13. 12 oz = __________ g **14.** 10 lb = __________ kg **15.** 6 in. = __________ cm

16. 5.5 ft = __________ m **17.** 2.5 gal = __________ L **18.** $2\frac{1}{4}$ mi = __________ km

19. 6.25 yd = __________ m **20.** 18 lb = __________ kg **21.** 15 fl oz = __________ L

22. $3\frac{1}{8}$ mi = __________ km **23.** $1\frac{3}{4}$ ft = __________ cm **24.** 2.5 qt = __________ L

25. 10 fl oz = __________ mL **26.** 15 qt = __________ L **27.** 220 mi = __________ km

8. 20 yd = __________ m **29.** 20.35 lb = __________kg **30.** 20 qt = __________ L

350.5 mi = __________ km **32.** 25 fl oz = __________ mL **33.** 4.5 lb = __________ kg

Name ______________________ Date ____________ Period ________

Converting Metric Units to Customary Units

You can use the following chart to convert customary units to metric units.

Customary Unit / Metric Unit
1 cm = 0.394 in.
1 m = 3.281 ft or 1.093 yd
1 km ≈ 0.621 mi
1 g ≈ 0.035 oz
1 kg = 2.205 lb g
1 mL ≈ 0.034 fl oz
1 L = 1.057 qt or 0.264 gal

Examples ***1*** **3 cm = ________ in.**

3 × 0.394 = 1.182 *1 cm ≈ 0.394 in., so multiply by 0.394.*

3 cm = 1.182 in.

2 **250 g = ________ oz**

250 × 0.035 = 8.75 *1 g ≈ 0.035 oz, so multiply by 0.035.*

250 g = 8.75 oz

3 **1.5 L = ________ qt**

1.5 × 1.057 = 1.5855 *1 L ≈ 1.057 qt, so multiply by 1.057.*

1.5 L = 1.5855 qt

Complete.

1. 5 cm = __________ in.

2. 787 g = __________ oz

3. 4 L = __________qt

4. 8 km = __________ mi

5. 2 L = __________ gal

6. 300 g = __________ oz

7. 155 mL = __________ fl oz

8. 9 km = __________ mi

9. 4 m = __________ yd

SKILL 46

Name ______________________ Date __________ Period _______

Converting Metric Units to Customary Units *(continued)*

Complete.

10. 3.5 km = __________ mi

11. 10 mL = __________ fl oz

12. 4.5 L = __________ gal

13. 7.5 m = __________ ft

14. 2.3 m = __________ yd

15. 3.5 L = __________ qt

16. 260 mL = __________ fl oz

17. 14 kg = __________ lb

18. 3.25 m = __________ ft

19. 24.5 km = __________ mi

20. 22 L = __________ gal

21. 45 g = __________ oz

22. 1.25 m = __________ ft

23. 12 kg = __________ lb

24. 14 L = __________ gal

25. 4.65 km = __________ mi

26. 4.8 cm = __________ in.

27. 8.5 L = __________ qt

28. 40 mL = __________ fl oz

29. 10.9 L = __________ gal

30. 280 km = __________ mi

31. 8 m = __________ yd

32. 15.35 kg = __________ lb

33. 10.5 L = __________ qt

6 cm = __________ in.

35. 15.5 m = __________ yd

36. 14 g = __________ oz

25 L = __________ qt

38. 50 kg = __________ lb

39. 2.8 m = __________ ft

SKILL 47

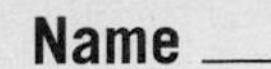

Name ______________________ Date ____________ Period ________

Adding and Converting Units of Time

Time
1 hour (hr) = 60 minutes (min)
1 minute (min) = 60 seconds

To add measures of time, add the seconds, add the minutes, and add the hours. Rename if necessary.

Example **Add 4 hours 25 minutes 40 seconds and 5 hours 30 minutes 25 seconds.**

4 h 25 min 40 s
+ 5 h 30 min 25 s

9 h 55 min 65 s = 9 h 56 min 5 s *Rename 65 s as 1 min 5 s.*

Rename each of the following.

1. 14 min 85 s = ____________ min 25 s

2. 8 h 65 min = 9 h ____________ min

3. 3 h 19 min 67 s = 3 h ____________ min 7 s

4. 6 h 68 min 25 s = ____________ h ____________ min 25 s

5. 7 h 105 min 15 s = ____________ h ____________ min 15 s

6. 4 h 99 min 80 s = ____________ h ____________ min ____________ s

7. 1 h 76 min 91 s = ____________ h ____________ min ____________ s

8. 7 h 88 min 60 s = ____________ h ____________ min ____________ s

SKILL 47

Name ________________ Date ________ Period ______

Adding and Converting Units of Time *(continued)*

Add. Rename if necessary.

9. 35 min 45 s
+ 12 min 12 s

10. 6 h 50 min
+ 3 h 17 min

11. 9 h 45 min 10 s
+ 3 h 30 min 50 s

12. 1 h 55 min 12 s
+ 3 h 25 min 34 s

13. 11 h 33 min 6 s
+ 5 h 36 min 29 s

14. 6 h 10 min 47 s
+ 2 h 51 min 28 s

15. 7 h 30 min 52 s
+ 3 h 45 min 40 s

16. 9 h 10 min 45 s
+ 3 h 55 min 30 s

An atlas gives average travel times. Use this information to answer Exercises 17-19.

Average Travel Times	
Baton Rouge to Mobile	4 h 40 min
Mobile to Tallahassee	5 h 50 min
Tallahassee to Jacksonville	3 h 35 min

17. What is the average travel time from Baton Rouge to Tallahassee going through Mobile?

18. What is the average travel time from Mobile to Jacksonville going through Tallahassee?

19. What is the average travel time from Baton Rouge to Jacksonville going through Mobile and Tallahassee?

)0. Wesley Paul set an age group record in the 1977 New York Marathon. He ran the race in 3 hours 31 seconds. He was 8 years old at the time. If he ran 2 hours 58 minutes 48 seconds in practice the day before the race, for how long did Wesley run on both days?

SKILL 48

Name ______________________ Date ____________ Period ________

Line Graphs

The diagram shows the parts of a graph.

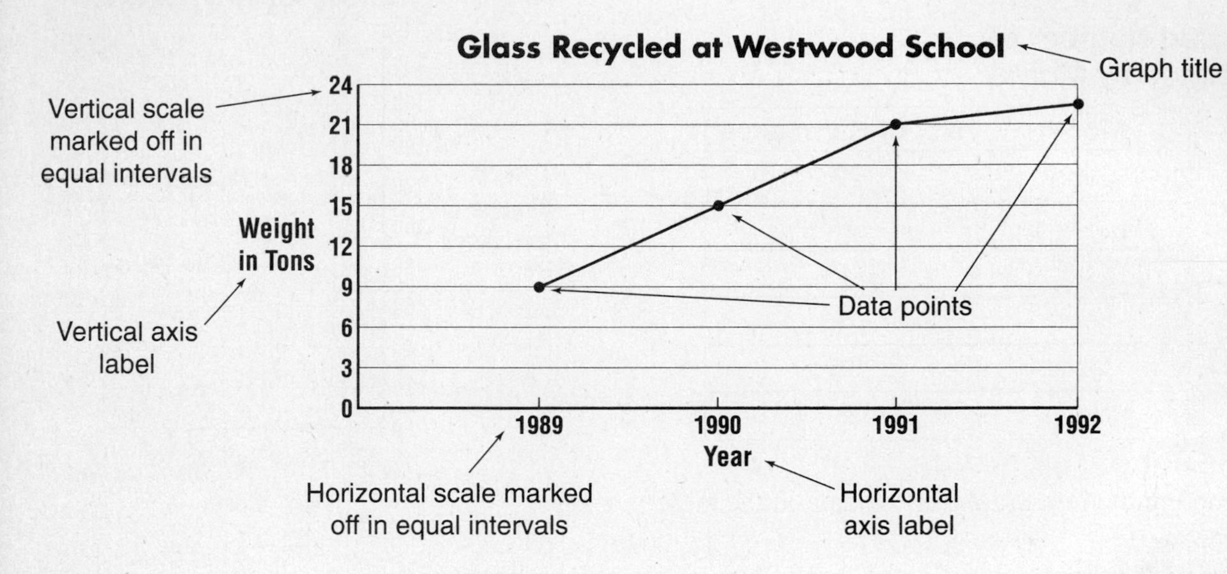

Solve.

1. Make a line graph for this set of data.

Number of Votes Expected	
Date	**Number of Votes**
3/15	18
3/30	11
4/15	15
4/30	9

2. Make a line graph for this set of data.

Evans Family Electric Bill	
Month	**Amount**
March	$129.90
April	$112.20
May	$105.00
June	$88.50

SKILL 48

Name ______________________ Date ____________ Period ________

Line Graphs *(continued)*

Refer to the following table for Exercises 1-2.

Recorded Number of Hurricanes by Month

Month	No. of Hurricanes
June	23
July	36
Aug.	149
Sept.	188
Oct.	95
Nov.	21

Number of Hurricanes

Number of Hurricanes: 0, 20, 40, 60, 80, 100, 120, 140, 160, 180, 200

June, July, Aug., Sept., Oct., Nov.

Month

3. Complete the line graph for the data in the table.

4. After which month does the number of hurricanes start to decrease?

Use the data in the table to complete the line graph.

5. **Temperatures on 2/15**

Time	Temperature
9:00 A.M.	32° F
11:00 A.M.	35° F
1:00 P.M.	38° F
3:00 P.M.	42° F
5:00 P.M.	39° F

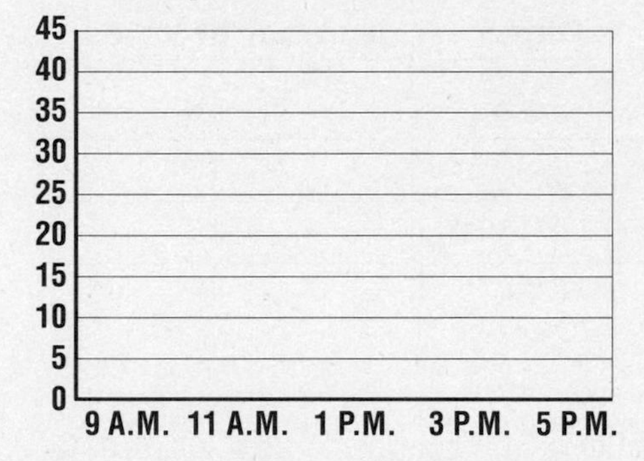

Solve. Use the line graph.

6. During which hour did the most rainfall occur?

How many inches of rain fell between 4 P.M. and 6 P.M.?

How many inches of rain fell between 3 P.M. and 8 P.M.?

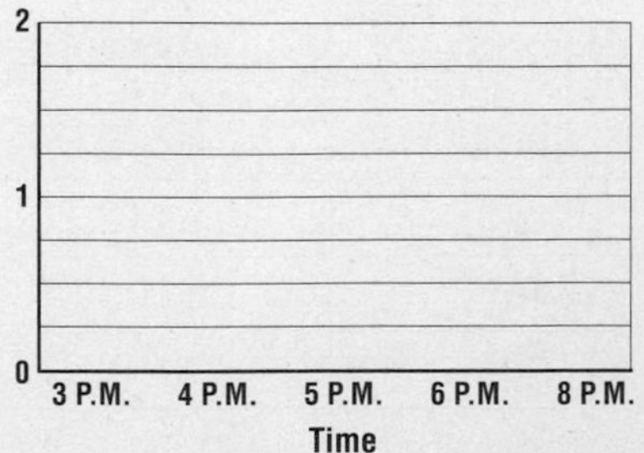

SKILL 49

Name ______________________ Date __________ Period ______

Histograms

A **histogram** uses bars to display numerical data that have been organized into equal intervals.

Example **The table shows the percent of people in several age groups who are not covered by health insurance. Make a histogram of the data.**

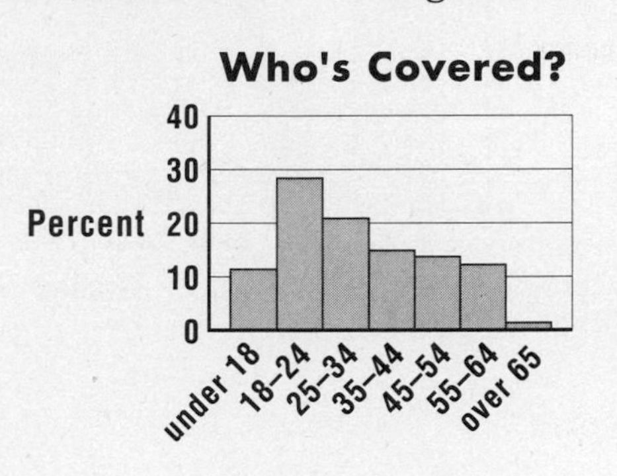

Age	Percent
under 18	12.4%
18-24	28.9%
25-34	20.9%
35-44	15.5%
45-54	14.0%
55-64	12.9%
over 65	1.2%

Make a histogram of the data below.

1.

Pieces of Junk Mail	Frequency
0-4	25
5-9	35
10-14	50
15-19	40
20-24	15

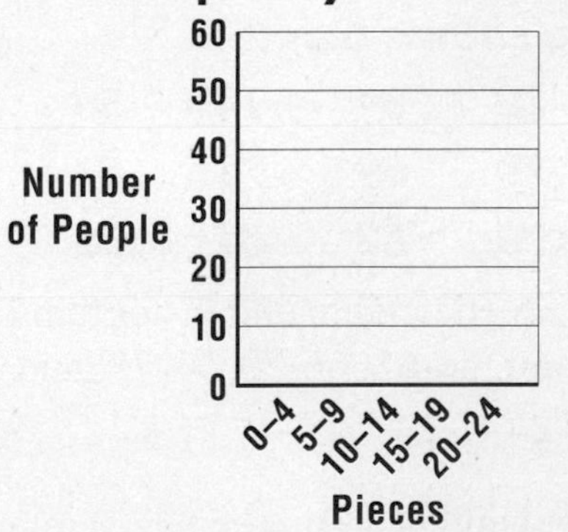

2.

Time Spent Surfing the Web (in hours per day)	Frequency
0-1	20
2-3	18
4-5	2
6-7	1

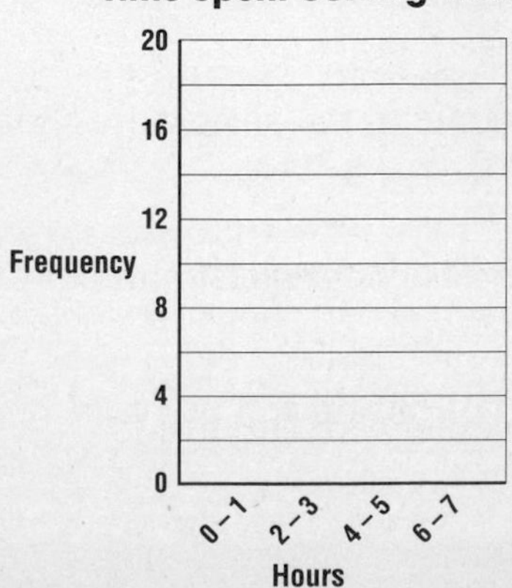

Name ____________________ Date __________ Period ______

Histograms *(continued)*

Use the histogram at the right to answer each question.

Algebra Test Scores

Frequency: 0, 1, 2, 3, 4, 5, 6

95–100 90–94 85–89 80–84 75–79

Grades

3. How many students took the algebra test?

4. Which grade has the most test scores?

5. Which grades have the same number of test scores?

6. How many more students earned 85–89 than earned 80–84?

7. Make a frequency table of the algebra scores.

A survey was taken that asked people their height in inches. The data are shown below.

68	*69*	*72*	*64*	*74*	*56*	*62*	*58*
69	*65*	*70*	*59*	*71*	*67*	*66*	*64*
73	*78*	*70*	*52*	*61*	*68*	*67*	*66*

8. Make a frequency table and histogram of the data. Use the intervals 51-55, 56-60, 61-65, 66-70, 71-75, and 76-80.

9. How many heights are in the 66-70 interval?

10. How many people in the survey are taller than 5 feet?

11. How many people in the survey are shorter than 5 feet?

'. What interval has the greatest number of heights?

How many people were surveyed?

SKILL 50

Name ______________________ Date __________ Period ______

Probability

The **probability** of an event is the ratio of the number of ways an event can occur to the number of possible outcomes.

$$\text{probability of an event} = \frac{\text{number of ways the event can occur}}{\text{number of possible outcomes}}$$

Example **On the spinner below, there are ten equally likely outcomes. Find the probability of spinning a number less than 5.**

Numbers less than 5 are 1, 2, 3, and 4.
There are 10 possible outcomes.

Probability of number less than 5 $= \frac{4}{10}$ or $\frac{2}{5}$.

The probability of spinning a number less than 5 is $\frac{2}{5}$.

A box of crayons contains 3 shades of red, 5 shades of blue, and 2 shades of green. If a child chooses a crayon at random, find the probability of choosing each of the following.

1. a green crayon

2. a red crayon

3. a blue crayon

4. a crayon that is *not* red

5. a red or blue crayon

6. a red or green crayon

SKILL 50

Name ______________________ Date __________ Period ______

Probability (continued)

A card is chosen at random from a deck of 52 cards. Find the probability of choosing each of the following.

7. a red card

8. the jack of diamonds

9. an ace

10. a black 10

11. a heart

12. *not* a club

A cooler contains 2 cans of grape juice, 3 cans of grapefruit juice, and 7 cans of orange juice. If a person chooses a can of juice at random, find the probability of choosing each of the following.

13. grapefruit juice

14. orange juice

15. grape juice

16. orange or grape juice

17. *not* orange juice

18. *not* grape juice

Businesses use statistical surveys to predict customers' future buying habits. A department store surveyed 200 customers on a Saturday in December to find out how much each customer spent on their visit to the store. Use the results at the right to answer Exercises 19–21.

Amount Spent	Number of Customers
Less than $2	14
$2–$4.99	36
$5–$9.99	42
$10–$19.99	32
$20–$49.99	32
$50–$99.99	22
$100 or more	22

'9. What is the probability that a customer will spend less than $2.00?

What is the probability that a customer will spend less than $10.00?

Vhat is the probability that a customer will spend between $20.00 and $100.00?